Automation in Clinical Microbiology

Editor

James H. Jorgensen, Ph.D.
Professor of Pathology, Microbiology, and
Medical Technology
University of Texas Health Science Center at San Antonio
and
Director
Microbial Pathology Laboratories
Medical Center Hospital
San Antonio, Texas

CRC Press
Taylor & Francis Group
Boca Raton London New York

CRC Press is an imprint of the
Taylor & Francis Group, an informa business

First published 1987 by CRC Press
Taylor & Francis Group
6000 Broken Sound Parkway NW, Suite 300
Boca Raton, FL 33487-2742

Reissued 2018 by CRC Press

Library of Congress Cataloging-in-Publication Data

Automation in clinical microbiology.

Includes bibliographies and index.
1. Diagnostic microbiology--Automation.
I. Jorgensen, James H. [DNLM: 1. Automation--instrumentation. 2. Computers. 3. Microbiological Technics--instrumentation. QW 26.5 A9387]
QR67.A98 1987 616'.01 86-32643
ISBN 0-8493-5108-1

A Library of Congress record exists under LC control number: 86032643

ISBN 13: 978-1-315-89089-0 (hbk)
ISBN 13: 978-1-351-06999-1 (ebk)

Visit the Taylor & Francis Web site at http://www.taylorandfrancis.com and the
CRC Press Web site at http://www.crcpress.com

THE EDITOR

James H. Jorgensen is Professor of Pathology, Microbiology, and Medical Technology at The University of Texas Health Science Center at San Antonio. He is also Director of the Microbial Pathology Laboratories at Medical Center Hospital, the principal teaching hospital of the Health Science Center.

Dr. Jorgensen was born in Dallas, Texas, in 1946. He received his Baccalaurate and Masters degrees from North Texas State University, and a Ph.D. in Microbiology from The University of Texas Medical Branch in Galveston. He joined the faculty of The University of Texas Medical School in San Antonio (now the Health Science Center) in 1973.

A member of several professional societies, he currently is Chairman-elect of the Clinical Division of the American Society for Microbiology, President of the Texas Infectious Diseases Society, and Program Director of the South Texas Association for Microbiology Professionals. He also is a member of the Committee for the Interscience Conference on Antimicrobial Agents and Chemotherapy, American Society for Microbiology President's Ad Hoc Commission on Clinical Microbiology, and the Subcommittee on Automation in Clinical Microbiology of the National Committee for Clinical Laboratory Standards.

Dr. Jorgensen has contributed numerous book chapters, reviews, editorials, and original articles on a variety of subjects in Clinical Microbiology, but he has especially emphasized the use of rapid and automated diagnostic procedures. He currently is a member of the editorial boards of *The Journal of Clinical Microbiology, Antimicrobial Agents and Chemotherapy,* and *Diagnostic Microbiology and Infectious Diseases.*

In recent years, he has been a frequent speaker regarding automation in microbiology at various international and national scientific gatherings.

CONTRIBUTORS

George E. Buck, Ph.D.
Associate Director
Clinical Microbiology Division and
Assistant Professor
Department of Pathology
University of Texas Medical Branch
Galveston, Texas

Richard L. Cohen, Ph.D.
Chief, Microbiology Section
Veterans Administration Medical Center
Dallas, Texas

Gary V. Doern, Ph.D.
Director
Department of Clinical Microbiology and
Associate Professor
Departments of Medicine and Molecular Genetics and Microbiology
University of Massachusetts Medical Center
Worcester, Massachusetts

Jack Goldstein, Ph.D.
Group Vice President
Ortho Diagnostic Systems, Inc.
Raritan, New Jersey

Harry R. Hill, M.D.
Director, Clinical Immunology Laboratory
Associated Regional and University Pathologist
Professor and Associate Chairman
Department of Pathology
University of Utah
Salt Lake City, Utah

James E. Johnson, Ph.D.
Chief, Serology Reference Laboratory
Veterans Administration Hospital and
Associate Professor
Department of Pathology
University of Kentucky Medical Center
Lexington, Kentucky

James H. Jorgensen, Ph.D.
Professor of Pathology, Microbiology, and Medical Technology
University of Texas Health Science Center at San Antonio
San Antonio, Texas

Lennart Larsson, Ph.D.
Research Chemist
Department of Medical Microbiology
University of Lund
Lund, Sweden

John M. Matsen, M.D.
Professor and Chairman
Department of Pathology
University of Utah School of Medicine
Salt Lake City, Utah

Frederick A. Meier, M.D.
Assistant Professor
Departments of Pathology and Medicine and
Medical Director
Clinical Microbiology Laboratory
Medical College of Virginia
Richmond, Virginia

Margie A. Morgan, Ph.D.
Chief Microbiologist
Department of Pathology and Lab Medicine
Cedars-Sinai Medical Center
Los Angeles, California

Patrick R. Murray, Ph.D.
Director
Clinical Microbiology Laboratory
Barnes Hospital and
Associate Professor of Pathology and Medicine
Washington University School of Medicine
St. Louis, Missouri

Marie Pezzlo, M.A.
Senior Supervisor
Medical Microbiology
University of California Irvine Medical Center
Orange, California

Michael A. Pfaller, M.D.
Director
Special Microbiology Laboratory and
Associate Professor
Department of Pathology
University of Iowa Hospitals and Clinics
Iowa City, Iowa

Glenn D. Roberts, Ph.D.
Consultant in Clinical Microbiology
Section of Clinical Microbiology
Department of Laboratory Medicine
Mayo Clinic and
Professor of Microbiology and Laboratory Medicine
Mayo Medical School
Rochester, Minnesota

Joseph L. Staneck, Ph.D.
Director
Division of Clinical Microbiology and
Associate Professor
Department of Pathology and Laboratory Medicine
University of Cincinnati Medical Center
Cincinnati, Ohio

TABLE OF CONTENTS

Introduction

Chapter 1

THE EVOLVING ROLE OF AUTOMATION IN CLINICAL MICROBIOLOGY

James H. Jorgensen

The laboratory diagnosis of an infectious disease is a complex, laborious, and frequently slow process compounded by the vast number of possibilities of infecting microorganisms, i.e., the fact that mixtures of microorganisms often exist in an infection, the complex nature of human body fluids and tissues, and the often uncertain status of the host's immunity. Much of contemporary clinical microbiology practice still is based on the pure culture techniques developed more than a century ago by Pasteur and the advent of agar-hardened growth media by Koch. Once bacteria could be cultivated in vitro, their tinctorial properties and preferred method of growth could be determined. Taxonomy of microorganisms became possible by differential growth characteristics on various complex or semi-defined media. Thus, determinations of the abilities of organisms to ferment sugars, degrade amino acids, grow at the expense of simple organic compounds, or produce unique metabolic end products have served as the backbone of bacterial taxonomy since the late 1800s.

In the late 1950s through the 1960s, miniaturization of otherwise traditional biochemical identification tests began first with smaller test tubes, and by the late 1960s, resulted in use of very small molded plastic reaction vessels of varying shapes. Somewhat later, paper strips or disks were impregnated with biochemical substrates providing further convenience and stability.

By the 1970s, taxonomic data had grown in complexity and sophistication to allow separation of microorganisms into hundreds of species based upon subtle differences in metabolism or antigenic composition. It then became important for numerous test results to be considered simultaneously with the help of computerized "data bases".

Not long after the discovery of antibiotics in the 1940s, it became important to make predictions on the likely effectiveness of a particular drug based upon in vitro testing. Antimicrobial susceptibility testing has been performed by a variety of techniques in the intervening 40 years. Various methods for standardizing in vitro tests have been extolled for the majority of this period.

Throughout these decades of development of clinical microbiology as a discipline, microbiologists became known by their colleagues in other disciplines as precise, highly skilled scientists who were fiercely independent and often shunned attempts to simplify the process of microbiological diagnosis. Before the advent of commercially prepared media and test kits, the art of microbiology was practiced in varying degrees of precision.

Despite the evolution of microbiology into a discipline of analytical precision, taxonomic detail, and now rapid result generation, microbiology has remained a laborious practice requiring highly skilled and experienced laboratorians. In contrast to other areas of the clinical laboratory which have long since adopted automated instrument methods of analysis, i.e., chemistry and hematology, microbiology has accepted automated instrumentation more slowly. This is in part due to the greater complexity of microbiological analyses, but is also due to a conservative and even skeptical attitude among microbiologists regarding nontraditional methodologies.

The development of the first truly automated microbiology instrument (Technicon Automated Susceptibility System, see Chapter 7) was considered such a commercial risk by its developers that it was never marketed. The speed, sensitivity, and labor-saving aspects of the instrument placed it more than a decade ahead of its time. While never marketed, it has served as a design goal for instruments which have been developed since its description.

At least four general categories of automated instruments for microbiology exist currently. There are several instruments which allow detection of growth of microorganisms (bacteria and fungi) directly from patients' specimens; some instruments provide a means of detection of microbial antigens or host antibody directed against microbes; instruments now exist which employ physical-chemical means of detecting the presence or growth of organisms; and lastly, a large number of instruments are available for organism identification and in vitro antimicrobial susceptibility testing.

Several reasons exist for use of automated instrumentation in microbiology. Sophisticated instruments should provide greater precision, shorten analysis time, reduce direct labor requirements, and potentially provide more accurate results at a reduced cost. A review of the chapters of this text will provide firm support for the first three of these goals. However, in several instances, the advantages of automation result in some compromise in accuracy as compared to the best currently available manual methods. Likewise, a major shortcoming of many current automated instruments is that an overall reduction in costs has not been attained. Often the price of performing a test by an instrument method exceeds the cost of a manual procedure. This is in contrast to the cost advantages of automated testing by chemistry and hematology autoanalyzers. Many clinical microbiologists have decided that they should add automation to their laboratories, but are now reluctant to do so in the rapidly changing climate of prospective payment plans for medical care by third-party payers. Thus, the current challenge to manufacturers is to design instruments which will provide more economical testing for clinical microbiology laboratories.

Microbiologists must also carefully consider the impact of new instrumentation on the laboratory staff and its external communication system. Careful studies of the impact of an instrument on work flow and staffing will be required for optimum utilization. Since many instrument methods provide test results much faster than manual test methods, the other two aspects of the overall scheme of laboratory testing must be reevaluated, i.e., the means of specimen acquisition and the laboratory's method of reporting results. An improvement in intralaboratory turnaround time through use of instrumentation may not lead to overall improved patient care unless transport of specimens and reporting of results can be likewise steamlined.

Selection of instrumentation must also include some predictions by the laboratory regarding future testing needs. Some instruments may be better suited to performance of a variety of tasks and methods and thus less likely to become prematurely obsolete. Changes in technology and the criteria which constitute contemporary medical practice are inevitable. A potential buyer should consider not only the current uses of an instrument but also the likely future commitment of its manufacturer to the field of microbiology. Is there provision for periodic updates for a particular instrument?

The chapters of this book describe numerous successful examples of automation in microbiology, e.g., radiometric detection of bacteremia, instruments for detection of bacteriuria, machines for organism identification and susceptibility testing, and automated antigen and antibody measurement systems. In addition, there are discussions of exciting but not yet proven methodologies such as chromatography, flow cytometry, and other applications of radiometry. There are also important discussions regarding improved means of data communication and ways to improve the clinician's use of test results. Lastly, there are candid assessments of the best and worst aspects of the current spectrum of automated instruments for microbiology. It is hoped that the reader of this volume will be left with a feeling of excitement at the possibilities that lie ahead for application of instrument techniques in the diagnosis of infectious diseases.

Instrument Approaches Currently in Use for Detection of Microorganisms in Body Fluids

Chapter 2

INSTRUMENTED APPROACHES TO PERFORMING BLOOD CULTURES

Gary V. Doern

TABLE OF CONTENTS

I. INTRODUCTION

Prior to 1970, detection of bacteremia and fungemia in the clinical laboratory was accomplished largely without the assistance of instrumentation. Blood for culture was inoculated either directly onto agar-containing plates or, more commonly, into bottles containing broth culture media. Direct inoculation of blood specimens onto plates is of limited value as a routine blood culture procedure due to the low sensitivity of this method.[1] Bacteremia in adults is usually characterized by relatively low concentrations of organisms in blood.[2] Since the total volume of blood which can be conveniently cultured with a direct plating method is small, the overall sensitivity of this technique is unacceptably low. In contrast, broth blood cultures permit sampling of larger volumes of blood, thus enhancing the likelihood of detecting bacteremia or fungemia. This approach, however, necessitates use of tedious and time-consuming manipulations for maximum rates of detection. Typically, according to a defined schedule, blood culture bottles must be examined macroscopically for evidence of growth, i.e., turbidity, hemolysis, gas formation, and/or colonies on the settled blood layer or sides of the bottles. In addition, at various intervals blood must be aseptically removed from blood culture bottles and subcultured blindly to solid media, as well as examined microscopically for bacteria and fungi with either the Gram or acridine orange stains.[3]

In 1970, Johnston Laboratories (Towson, Md.) introduced an alternative approach to performing blood cultures which is based on the radiometric detection of bacteremia and fungemia through use of an instrument, the Bactac® blood culture analyzer.[4] The Bactec® system utilizes blood specimens inoculated into broth, as is the case with conventional nonradiometric broth blood cultures; however, the detection of microorganisms when present in blood culture bottles is accomplished with the Bactec® instrument. As a result, the necessity of examining blood culture broth microscopically and by blind subculture is eliminated.[5-7] Since the inception of the Bactec® radiometric blood culture system, several other instrumented approaches to the detection of bacteremia and fungemia have been developed. These include techniques based on microcalorimetry,[8] pH,[9] monitoring of electrical potentials,[10] colorimetry,[11] flow cytometry,[12] filtration-bioluminescence,[13] electrical impedence measurements,[14] and infrared detection of CO_2.[15] The intent of this chapter is to examine the utility of these instrumented approaches to performing blood cultures.

II. THE BACTEC® SYSTEM

The Bactec® blood culture system is based on the measurement of ^{14}C-labeled CO_2 in the head-space air of blood culture bottles containing tryptic soy broth, various growth supplements, and ^{14}C-labeled growth substrates. Currently, Johnston Laboratories markets five different media for use with the Bactec® radiometric blood culture system: a nonhypertonic aerobic medium (6B), a hypertonic aerobic medium (8B), an anaerobic medium (7D), a resin-containing aerobic medium (16B), and a resin-containing anaerobic medium (17D). In addition, American Micro Scan (Mahwah, N.J.) recently introduced two radiometric blood culture media said to be compatible with the Bactec® system, an aerobic medium (A) and an anaerobic medium (B). Blood specimens are inoculated into blood culture bottles containing the desired medium, incubated at 35°C and examined periodically for the production of ^{14}C-labeled CO_2 using the Bactec® blood culture analyzer.

The instrument samples blood culture bottles aseptically by aspirating a portion of the head-space air into an ionization chamber in the machine. The amount of $^{14}CO_2$ present is quantitated, digitalized on an arbitrary linear scale, and displayed as a growth index both on an instrument display screen as well as a paper tape. A specific growth index may be taken as evidence of the presence of microorganisms in the blood culture broth when it exceeds an established background index. The volume of head-space air sampled to obtain each reading is automatically replenished with gas of known composition, maintained in gas tank reservoirs. In this manner, the head-space atmosphere of both aerobic and anaerobic blood culture bottles can be maintained at a constant.

During the last 15 years, there have been nearly 200 published investigations pertaining to use of the Bactec® blood culture system. Based on the results of these studies, and assuming that optimal detection of bacteremia necessitates sampling of at least 20 mℓ of blood specimen per culture[16,17] with a dilution of at least 1:5 when blood specimen is inoculated into broth medium,[18,19] the following observations seem to apply to use of the Bactec® system. Maximum rates of detection will be achieved when 20 mℓ blood specimens are distributed in equal 6.6 mℓ aliquots into aerobic (6B), aerobic hypertonic (8B), and anaerobic (7D) Bactec® blood culture media. In patients receiving antimicrobial therapy at the time blood specimens are obtained, resin-containing aerobic (16B) and anaerobic media (17D) should be substituted for 6B and 7D media, respectively. At the present time, given the absence of any published studies documenting the efficacy of Micro Scan® radiometric media A and B, these media cannot be recommended for use with the Bactec® system. Obviously, if in the future these media are shown to have performance characteristics comparable or superior to Bactec® media, they could be considered as substitutes for Bactec® aerobic (6B) and anaerobic (7D) media.

All blood culture bottles should be incubated at 35°C for at least 7 days. During the first 24 to 48 hr of incubation, aerobic bottles should be agitated on the Bactec® shaker at 260 rpm. Using the Bactec® blood culture analyzer, aerobic bottles should be examined for the presence of growth at least twice, preferably 3 or 4 times daily during the first 48 hr of incubation and at some interval thereafter ensuring that a terminal instrument reading is performed prior to being discarded as negative. Anaerobic bottles need not be agitated. They should be examined with the Bactec® instrument once on days one and two of incubation and at some interval thereafter. Again, a terminal reading should be obtained before discarding anaerobic bottles as negative. It should be emphasized that all blood culture bottles should be carefully examined for macroscopic evidence of growth before being placed on the Bactec® instrument. Macroscopically positive bottles should not be examined on the machine; they should be considered positive and processed accordingly.

Despite the fact that there have been numerous published evaluations of the Bactec® blood culture system during the last 15 years, surprisingly little data exists upon which to base meaningful comparisons of this system in its current form with conventional nonradiometric blood culture procedures.[20,21] Such comparisons are best predicated on objective studies which utilize clinical specimens analyzed in parallel with equal volumes of blood processed simultaneously in the blood culture systems being compared. Furthermore, it is essential that such studies be performed with currently available media formulations. The current nonhypertonic and hypertonic Bactec® aerobic media are different from initial formulations. The Bactec® anaerobic medium has been changed three times. Obviously, an evaluation performed with earlier media formulations has no value in assessing the utility of currently available media formulations. Unfortunately, there have been few published systematic comparisons of the Bactec® system with conventional nonradiometric broth procedures which have been performed in a completely objective manner and which have utilized the formulations of the Bactec® blood culture media that are presently commercially available. For this reason, it is difficult to state with certainty precisely what the detection sensitivity of the Bactec® system is in comparison to conventional nonradiometric broth blood culture procedures. Based on what information does exist, however, the following two statements are probably true. When used as described previously, overall bacteremia and fungemia detection rates achievable with the Bactec® system would be at least equivalent to those obtained with conventional nonradiometric blood culture methods. Furthermore, up to 75% of all positive blood cultures will be detected within 24 hr of inoculation.

The principal advantage of the Bactec® blood culture system vs. conventional nonradiometric broth blood culture methods is the labor savings associated with instrument detection of positive cultures. With the Bactec® system, blood cultures need not be subjected to labor-intensive, time-consuming manipulations such as blind microscopic examination and subculture as a means for detecting the presence of microorganisms.[5-7] Rather, blood culture bottles are examined with an instrument for the presence of $^{14}CO_2$ in the head-space atmosphere as evidence of positivity. The only labor involved is transferring blood culture bottles from the incubator onto the machine. With the Bactec® 460 blood culture analyzer, 60 bottles may be loaded at one time. Approximately 1 min is required for machine sampling of each bottle. The work-load value assigned to a single blood culture bottle processed with the Bactec® 460 instrument by the College of American Pathologists is 4.1 units.[22] In contrast, 5.0 units per bottle are assigned for processing conventional nonradiometric blood cultures. This value includes two blind subcultures. Blind microscopic examination adds 2.2 units to the total work-load factor. Assuming a nonradiometric blood culture bottle is blind Gram stained once and blind subcultured twice, the total work-load factor would be 7.2 units. In other words, use of the Bactec® blood culture system represents a labor savings of approximately 43% vs. conventional nonradiometric methods.

Another advantage of the Bactec® system is the possibility for automatic storage and tabulation of the results of large numbers of blood cultures via direct linkage with a data management system. A data management computer system specifically designed for use with the Bactec® blood culture system is presently commercially available from Johnston Laboratories. This system, which is interfaced directly with the Bactec® blood culture analyzer, permits easy determination of rates of detection of significant isolates, length of time to detection, contamination rates, frequency of isolation of specific organisms and rates of detection of bacteremia or fungemia sorted by time and hospital location. Alternatively, the Bactec® blood culture analyzer could potentially be interfaced directly with laboratory-based computer systems for purposes of similar data analysis.

The primary disadvantage of the Bactec® blood culture system is the necessity of using blood culture media which are compatible with the system. In practical terms, this means that users of the Bactec® system must employ radiometric blood culture broth. As mentioned above, in the case of the Bactec® radiometric media, formulations have been changed frequently by the manufacturer. Since the intent of such media formulation changes is presumably to improve performance of the system, it is not unreasonable to expect that newer media formulations will enhance the efficacy of the system as a means for detecting bacteremia and fungemia. Unfortunately, at the time new media formulations are introduced, there often exists insufficient information upon which to base this assertion. The user, however, is "stuck" with the new media. Indeed, published objective comparisons of new formulations of Bactec® radiometric media with older versions do not exist. In a very real sense, the user is thus totally dependent on the manufacturer's internal evaluations.

Another potential disadvantage of the Bactec® blood culture system — for that matter, a potential disadvantage of any instrumented blood culture procedure — is the reliance on a machine for detecting positive blood cultures. What happens when the machine breaks down? In the case of the Bactec® system, except for inconvenience, this poses few problems. Radiometric blood culture media are broth based and, as a result, may be processed in the same manner as conventional nonradiometric blood cultures. Furthermore, it should be emphasized that when the manufacturer's recommendations concerning preventive maintenance are followed, the Bactec® 460 blood culture analyzer is an extremely reliable instrument. During a 3-year period in the author's laboratory, despite nearly constant use, this instrument was inoperative for a total of 16 hr.[30]

Other concerns pertaining to use of the Bactec® system include the risk of contamination of bottles with the instrument, the hazards of using and disposal of radioactive material, and the small volumes of blood which can be cultured in blood culture bottles compatible with the instrument.

Cross-contamination of negative blood culture bottles by the instrument has been reported as a cause of pseudobacteremia by various users of the Bactec® system.[23,24] Strict adherence to the manufacturer's instructions regarding preventive maintenance and performance checks on the instrument's sampling device sterilizer, tilting of aerobic bottles incubated on the shaker prior to machine reading to remove broth trapped under the bottle's rubber septa, and avoidance of testing macroscopically positive bottles on the instrument will effectively eliminate the problems of cross-contamination.

The amount and nature of radioactive material present in radiometric blood culture bottles (i.e., 2.0 μCi of ^{14}C per bottle) poses no risk to laboratory workers when contained in blood culture bottles. Spills of radiometric broth or contact to skin also pose no undue risk; however, appropriate precautions should be exercised when dealing with such incidents, i.e., thorough cleansing and rinsing of the contaminated surface with adequate amounts of water. The contents of used bottles may be disposed of without the need for precautionary measures such as biohazardous radioactive containment.

Finally, the issue of the relatively small volume of blood that can be processed in a single radiometric blood culture bottle is not a limitation of the system when used as described above. Each bottle contains 30 mℓ of broth medium. The 3 bottle system advocated previously permits sampling of 20 mℓ of blood specimen per blood culture while maintaining a dilution of >1:5, i.e., 6.6 mℓ blood specimen per 30 mℓ of broth.

In summary, the Bactec® system represents the first successful attempt at automating blood cultures. Through the effective use of instrumentation, this system permits performance of blood cultures with a significant savings of technologists' time and with the possibility for convenient and extensive data analysis. The rate of detection of bacteremia and fungemia as well as the length of time to detection achievable with the Bactec® blood culture system is probably at least equivalent to that achievable with conventional nonradiometric broth procedures.

III. OTHER INSTRUMENTED APPROACHES TO BLOOD CULTURES

A variety of other instrumented approaches to performing blood cultures have developed. These include techniques based on microcalorimetry,[8] pH,[9] monitoring of electrical potentials,[10] colorimetry,[11] flow cytometry,[12] filtration-bioluminescence,[13] electrical impedence measurements,[14] and infrared light detection of CO_2.[15] Of these, the last two may have potential utility as means for the routine detection of bacteremia in the clinical laboratory and therefore will be discussed further.

Growth of microorganisms in broth is accompanied by proportional changes in electrical impedence.[14,15,26] Taking advantage of this fact and utilizing an instrument which measures electrical impedence, the Bactometer® 32 (Bactomatic, Inc., Palo Alto, Calif.), Hadley and Senyk found that 10^6 to 10^7 bacteria per milliliter of blood culture medium could be detected by measuring changes in electrical impedence.[14] In a parallel clinical laboratory evaluation, the total number of positive blood cultures detected by electrical impedence was essentially comparable to that achieved with a conventional nonradiometric broth blood culture procedure.[17] Most microorganisms were detected more rapidly by electrical impedence. Similar observations were made when electrical impedence was compared to a radiometric blood culture method in a limited clinical laboratory evaluation.[28] Subsequently, it was shown that overall rates of detection increased and the length of time to detection decreased when electrical impedence was combined with lysis-filtration.[29] Blood specimens were first lysed and then processed through a membrane filter. The filter, containing trapped microorganisms, was placed in a broth medium and the electrical impedence of the broth was monitored continuously with the Bactometer® instrument. The limit of sensitivity of this method was found to be approximately 5×10^4 microorganisms per mℓ of broth.

In 1984, Johnston Laboratories introduced an instrumented blood culture system based on the detection of CO_2 in the head-space atmosphere of blood culture bottles using infrared spectrophotometry. This system is predicated on the observation that infrared radiation is absorbed by CO_2. Briefly, by means of a Bactec® 660 blood culture instrument, infrared light is passed through a sample cell containing gas aspirated from the head-space atmosphere of a blood culture bottle. The amount of infrared light transmitted through the sample cell is registered on a detector and is inversely proportional to the amount of CO_2 in the head space. Values obtained with blood culture bottles are normalized against the amount of CO_2 contained in the head-space atmosphere due to the incubation gas mixture and compared to values obtained with a reference sample cell. The final reading is presented in the form of a digital growth value. At the present time, four blood culture media compatible with this system are commercially available: a nonhypertonic aerobic medium (NR6A), a hypertonic aerobic medium (NR8A), an anaerobic medium (NR7A) and a resin-containing anaerobic medium (NR17A). Each contains 30 mℓ of tryptic soy broth plus various growth supplements.

In a clinical laboratory comparison of this system with the Bactec® radiometric blood culture procedure, among a total of 5200 paired blood cultures, no difference in overall rates of detection or length of time to positivity were noted.[15] The infrared detection system, however, was found to have a faster cycle time, that is, blood culture bottles were processed more rapidly by the machine, it was easier to use, and eliminated the necessity of using blood culture media which contain radioactive material.

What role measurement of changes in electrical impedence or infrared spectrophotometry will play in the future as methods for performing blood cultures routinely in the clinical laboratory awaits further investigation. It is possible, however, that these or other instrumented approaches could have a considerable impact on the laboratory diagnosis of bacteremia and fungemia.

REFERENCES

1. DuPont, H. L. and Spink, W. W., Infections due to gram-negative organisms: an analysis of 860 patients with bacteremia at the University of Minnesota Medical Center, 1958—1966, *Medicine*, 48, 307, 1969.
2. Washington, J. A., II, Ed., Characteristics of bacteremia relevant to its laboratory diagnosis, in *The Detection of Septicemia*, CRC Press, West Palm Beach, Fla., 1978, 28.
3. Reller, L. B., Murray, P. R., and MacLowry, J. D., Blood cultures II, in *Cumulative Techniques and Procedures in Clinical Microbiology*, Washington, J. A., II, Ed., American Society for Microbiology, Washington, D.C., 1982, 6.
4. Deland, F. H. and Wagner, H. N., Jr., Automated radiometric detection of bacterial growth in blood cultures, *J. Lab. Clin. Med.*, 75, 529, 1970.
5. Burdash, N. M., Manos, J. P., Bannister, E. R., and Wellborn, A. L., Acridine organge staining and radiometric detection of microorganisms in blood cultures, *J. Clin. Microbiol.*, 17, 463, 1983.

6. Plorde, J. J., Carlson, L. G., and Dan, M. E., Lack of clinical relevance in routine final subcultures of radiometrically negative Bactec blood culture vials, *Am. J. Clin. Pathol.*, 78, 753, 1982.
7. McLaughlin, J. C., Evers, J. L., and Officer, J. L., Lack of requirement for blind subcultures of Bactec blood culture media, *J. Clin. Microbiol.*, 14, 567, 1981.
8. Ripa, K. T., Mardh, P.-A., Hovelius, B., and Ljungholm, K., Microcolorimetry as a tool for evaluation of blood culture media, *J. Clin. Microbiol.*, 5, 393, 1977.
9. King, G. S., Niles, A. C., and Murray, P. R., Evaluation of detection of positive blood cultures by pH changes, *J. Clin. Microbiol.*, 14, 591, 1981.
10. Holland, R. L., Cooper, B. H., Helgeson, N. G. P., and McCracken, A. W., Automated detection of microbial growth in blood cultures by using stainless-steel electrodes, *J. Clin. Microbiol.*, 12, 180, 1980.
11. Wallis, C. and Melnick, J. L., Rapid colorimetric method for the detection of microorganisms in blood culture, *J. Clin. Microbiol.*, 21, 505, 1985.
12. Mansour, J. D., Robson, J. A., and Schulte, T. H., The rapid detection of bacteria in seeded blood culture using flow cytometry: detection at <100 cfu/mℓ in less than two hours, in *Abstr. Annu. Meet. Am. Soc. Microbiol.*, American Society for Microbiology, Washington, D.C., 1985, 353.
13. Boheme, K., Ramey, N., and Tsai, T. L. S., Rapid detection of bacteria in blood cultures by filtration and bioluminescence, in *Abstr. Annu. Meet. Am. Soc. Microbiol.*, American Society for Microbiology, Washington, D.C., 1985, 353.
14. Hadley, W. K. and Senyk, G., Early detection of microbiol metabolism and growth by measurement of electrical impedence, in *Microbiology*, Schlessinger, D., Ed., American Society for Microbiology, Washington, D.C., 1975, 18.
15. Jungkind, D. and Millan, J., Clinical comparison of a new automated infrared blood culture system with the Bactec 460 system, in *Abstr. Interscience Conf. Antimicrob. Agents Chemother.*, American Society for Microbiology, Washington, D.C., 1984, 279.
16. Illstrup, D. M. and Washington, J. A., II, Importance of volume of blood cultured in the detection of bacteremia and fungemia, *Diagn. Microbiol. Infect. Dis.*, 1, 107, 1983.
17. Tenney, J. H., Reller, L. B., Mirrett, S., Wang, W.-L. L., and Weinstein, M. P., Controlled evaluation of the volume of blood cultured in detection of bacteremia and fungemia, *J. Clin. Microbiol.*, 15, 558, 1982.
18. Salventi, J. F., Davies, T. A., Randall, E. I., Whitaker, S., and Waters, J. R., Effect of dilution on recovery of organisms from clinical blood cultures in medium containing sodium polyanethol sulfonate, *J. Clin. Microbiol.*, 9, 248, 1979.
19. Reimer, L. G., Reller, L. B., Mirrett, S., Wang, W. L., and Cox, R. L., Controlled evaluation of hypertonic sucrose medium at 1:5 ratio of blood to broth for detection of bacteremia and fungemia in supplemented peptone broth, *J. Clin. Microbiol.*, 17, 1045, 1983.
20. Strand, C. L., Jones, M. C., and Daniel, W. W., Comparison of a radiometric and conventional blood culture system: efficacy of recovery, speed of recovery, cost and technical time, *Lab. Med.*, 11, 41, 1980.
21. Colmer, J. and Sodeman, T. M., Bactec automated blood cultures, *Lab. Med.*, 15, 534, 1984.
22. Workload Recording Committee, Laboratory Sections — Special Directions, in *Workload Recording Method*, College of American Pathologists, Skokie, Ill., 1984, chap. 4.
23. Griffin, M. R., Miller, A. D., and Davis, A. C., Blood culture cross contamination associated with a radiometric analyzer, *J. Clin. Microbiol.*, 15, 567, 1982.
24. John, J. F., Jr. and Bannister, E. R., Pseudobacteremia, *Infect. Contr.*, 5, 69, 1984.
25. Cady, P., Detection of microorganisms by electric impedence, in *New Approaches to the Identification of Microorganisms*, Heden, C. G. and Illeni, T., Eds., John Wiley & Sons, New York, 1975, 73.
26. Ur, A. and Brown, D. F. J., Electrical impedence measurement of bacterial growth, *J. Med. Microbiol.*, 8, 19, 1975.
27. Hadley, W. K. and Kazenka, W., Detection of bacteremia by electrical impedence, in *Abstr. Annu. Meet. Am. Soc. Microbiol.*, American Society for Microbiology, Washington, D.C., 1976, Abstr. C69.
28. Specter, S., Throm, R., Strauss, R., and Friedman, H., Rapid detection of bacterial growth in blood samples by a continuous-monitoring electrical impedence apparatus, *J. Clin. Microbiol.*, 6, 489, 1977.
29. Zierdt, C. H., Kagan, R. L., and MacLowry, J. D., Detection of bacteremia by lysis-filtration and electrical impedence monitoring, *J. Clin. Microbiol.*, 5, 46, 1977.
30. Doern, G. V., unpublished observations.

Chapter 3

INSTRUMENT METHODS FOR DETECTION OF BACTERIURIA

Marie Pezzlo

TABLE OF CONTENTS

I. INTRODUCTION

Urine specimens represent the most common specimen type sent to the laboratory for culture. The diagnostic microbiology laboratory faces the challenge of isolating the causative agent of urinary infections and reporting the results to the physician as rapidly as possible. Several variables must be taken into account when processing these specimens. The diagnostic criterion of a single species at a concentration of $\geqslant 10^5$ colony-forming units (CFU)/mℓ in a clean-voided urine sample from an asymptomatic patient continues to be a generally accepted definition of significant bacteriuria.[1,2] More recently, it has been reported that as few as 10^2 CFU/mℓ of coliforms and the presence of pyuria are significant findings in young symptomatic females.[3,4]

Many methods are available for screening urine specimens for bacteriuria. These include microscopic, chemical, culture, and automated techniques. Automated methods have been developed largely as a result of the need for rapid screening techniques. The purpose of rapid screening is to provide information to the physician in a timely manner which in turn leads to more prompt care of the patient. Results from instrumentation are available within 2 min to 13 hr, depending upon which method is used. Automated methods have been developed to meet the increased interest in rapid screening for bacteriuria. The automated methods in clinical use as well as those under investigation will be reviewed. These methods include bioluminescence, colorimetric filtration, electrical impedance, electrochemical, flow cytometry, laser nephelometry, microcalorimetry, particle size distribution analysis, photometry (nephelometry), and radiometry. Each of these methods will be reviewed with regard to its application in the diagnostic microbiology laboratory, detection times, and predictive values. The cost for testing will be discussed if available.

II. AUTOMATED URINE SCREENING — CLINICAL APPLICATIONS

The automated methods which are available for use in the diagnostic laboratory may be divided into three groups based on the principles of detection and their detection times. The most rapid method is colorimetric filtration since results are available within 2 min. This system is based on the presence of bacteria in urine, rather than growth. If bacteria are present in urine, they are trapped on a filter card and their presence is displayed by a pink color as result of staining with safranin dye. This method may also be used to detect pyuria. The instrument used for this assay is the Bac-T-Screen.®

The second group of automated bacteriuria screening methods utilizes the principle of bioluminescence for detection. Bacterial adenosine triphosphate (ATP) is assayed directly or following the lysis of somatic cell ATP. Results are available within 2 min to 1 hr. Three luminometers are available for performing bioluminescence assays in the diagnostic laboratory. These include the Lumac® system, the Microscreen system and the Turner® bacterial analysis kit.

The third group requires growth of the organism prior to detection and, therefore, results are somewhat delayed. Detection of growth is based on changes in light transmission photometry and results are available within 30 min to 13 hr. The three instruments which detect bacterial growth in urine are the Autobac® system, the Auto-Microbic® system, and the Avantage/MS®-2 system.

Table 1
COMPARISON OF RAPID[a] AUTOMATED BACTERIURIA SCREENS

	Sensitivity (%) CFU/mℓ			PVN[b] (%) CFU/mℓ			Cost per test[c] ($)	Instrument cost ($)	Detection time (min)
Method	10^2	10^4	10^5	10^2	10^4	10^5			
Bac-T-Screen®	79[d]	93	>92	93[d]		97	1.00	5000	2
Lumac®		>85	>92		95	99	1.10	8000	30
Monoscreen		>85[e]			96[e]		1.00	6500	15
Turner®		86	92		89	97	0.50—1.00	8000	1—60

[a] Less than 1 hr.
[b] Predictive value negative.
[c] Reagent cost only.
[d] Includes pyuria (>10 WBC/mm^3).
[e] $\geq 5 \times 10^4$ CFU/mℓ.

III. DIRECT DETECTION SYSTEMS

A. Colorimetric Filtration

The Bac-T-Screen® (Marion Laboratories, Inc., Kansas City, Mo.) is a 2-min urine screen in which 1 mℓ of well-mixed urine and 3 mℓ of urine diluent (14.5% acetic acid) are suctioned through a filter card followed by the addition and filtration of 3 mℓ of safranin O dye and 3 mℓ of 2.4% acetic acid decolorizer. The color intensity remaining is coded using a color guide on the filter card. A positive test is one which gives a pink residual color on the filter card of ≥1+. A card reader is provided to aid with interpretation of questionable results. Final results are available within 2 min.

The Bac-T-Screen® is a bacteriuria screening test which does not require growth for detection, but compares favorably to other screening methods.[5-10] The initial investigation of the Bac-T-Screen® reported an overall correlation of 98% when compared to a reference culture method.[10] In this study, $\geq 10^5$ CFU/mℓ was defined as a positive specimen and no data were available regarding detection of low-count bacteriuria. Additional studies[5-9] reported on the accuracy after modification of the instrument (Table 1). When the Bac-T-Screen® was compared to a growth detection method, the sensitivity was higher, and it detected all isolates of yeast which were missed by the photometric screen.[7] The Bac-T-Screen® has also been compared to the Gram stain since both methods are growth independent and provide results within 2 min.[9] The sensitivity and negative predictive value for both methods were similar; however, the Bac-T-Screen® eliminated observer variability.

As a result of the significance of low-count bacteriuria and the presence of pyuria in the young symptomatic female, the Bac-T-Screen® has been evaluated for its ability to detect low-level bacteriuria ($\geq 10^2$ CFU/mℓ) and significant pyuria (>10 WBC/mm^3).[11,12] When >10 WBC/mm^3 were included in the data analyses, the sensitivity and negative predictive values were increased (Table 1).[12] When evaluating the ability of the Bac-T-Screen® to detect significant pyuria (>10 WBC/mm^3), the sensitivity and negative predictive value was 93 and 96%.[11] The Bac-T-Screen® has the advantage of detecting bacteriuria and significant pyuria. A modification in the vacuum pump (Model 402) has eliminated a majority of the cloggers.

A disadvantage of the Bac-T-Screen® has been the uninterpretable specimens.[5,6,9] These represent specimens which clog the filter or leave a residual yellow to brown pigment on the card. These uninterpretable specimens result from the presence of large numbers of bacteria and white and red blood cells. For this reason, these specimens should be interpreted as positive based upon their correlation to bacteriuria and/or pyuria. A modification in the vaccuum pump (Model 1402) has eliminated a majority of the cloggers.

The cost of the Bac-T-Screen® is comparable to other automated methods,[7] and it has the advantage of being a 2-min test. It also has the advantage of detecting low-level bacteriuria and significant pyuria. The instrument is simple to operate and requires a small amount of laboratory space. The ease of operation allows it to be used in an outpatient clinic or physician's office.

B. Bioluminescence

The application of bioluminescence for use with clinical specimens was first described in 1944;[13] however, this technology has only recently been introduced in the diagnostic laboratory. The method is based on the bioluminescent reaction of adenosine-5′-triphosphate (ATP) with luciferin and luciferase, which is measured in a luminometer. Measurements may include both somatic and bacterial ATP or bacterial ATP alone after the removal of nonbacterial ATP. Results are displayed as relative light units (RLU).

Somatic cell ATP may be extracted with a nucleotide releasing agent and destroyed with an apyrase. The apyrase is then removed and bacterial ATP is extracted. Released bacterial ATP is detected by adding firefly luciferin-luciferase which results in the production of light which is proportional to the released bacterial ATP. This principle has been applied to screening for bacteriuria.

Earlier studies employing bioluminescence found the procedure to be time-consuming and the sensitivity varied greatly depending on the ATP concentration limit for interpretation.[14-16] Modifications of earlier instrumentation and simplification of the procedure have made bioluminescence a viable bacteriuria screening method. Three luminometers are available for urine screening. Each differs slightly in test performance and detection time. Two of the systems measure bacterial ATP and one measures ATP from both somatic and bacterial cells. Unlike earlier methods based on the ATP assay, these systems utilize purified reagents, more sensitive instrumentation and simplified technology.

The Lumac/3M® Bacteriuria Screening Kit (3M Co., St. Paul, Mn.) is used with the Lumac/3M® M2010 Biocounter to determine the presence of bacteria in urine by measuring bacterial ATP. Nucleotide-releasing somase is added to 25 $\mu\ell$ of urine and allowed to incubate for 25 min at 35°C. The sample is then transferred to the Biocounter and nucleotide-releasing reagent for bacterial cells is added followed by luciferin-luciferase. Light emitted is displayed as RLU.

The Lumac® system has a sensitivity of >92% using ≥200 RLUs as a positive result.[17-21] This reading has been selected since it seems to give the least number of false-negative and false-positive results when significant bacteriuria has been defined as $\geq 10^4$ CFU/mℓ.

The threshold bioluminescence levels of 500, 800, and 2000 RLU have also been used.[20-22] Although there are fewer false-positive results when the RLU is increased, the sensitivity of the test is decreased. The percentage of false-positives ranges from 14 to 25% when using 200 RLU as the cutoff level with a sensitivity of >92% at $\geq 10^5$ CFU/ℓ and >85% at $\geq 10^4$ CFU/mℓ (Table 1).[17,18] Results are available within 30 min when the assay is performed by the Lumac® Biocounter.

The cost per test given by the manufacturer is approximately $1.10 and the Lumac® Biocounter M-2010 is $8000. When reagent cost, technical time, and cost of plating true and false positives are considered, the average cost per specimen is approximately $2.50.[17]

The bioluminescence assay may also be performed by the Microscreen system (Analytical Luminescence Laboratory, San Diego, Calif.) which consists of the Monolight

500 luminometer and the Micromate® 501 dispenser/diluter. To obtain results with this system, 25 $\mu\ell$ of urine is placed into a precoated assay cuvette and allowed to incubate at room temperature for 10 min. During this incubation time, the released somatic ATP is being converted to AMP and therefore will not interfere with the assay of bacterial ATP. The cuvette is then placed into the Monolight 500 luminometer and bacterial reagent and luciferin-luciferase enzyme is added automatically. If sufficient bacterial ATP is present, light is produced and measured by the luminometer. A reading of ≥200 RLU is considered positive by this method also. Results performed by this assay are available within 15 min.

In a study of 976 urine specimens,[24] the overall detection rate was 85.3% and the false-positive rate was 21.8% when $\geq 5 \times 10^4$ CFU/mℓ was considered positive. However, when false-positive and false-negative results were further examined, the presence of low colony counts ($<5 \times 10^4$ CFU/mℓ) residual antimicrobial activity and antibody-coated bacteria were found in some specimens. Results from the Monolight luminometer and Lumac® Biocounter were compared and found to be similar.[17] When the evaluation was performed, the Monolight procedure required more dilution steps and centrifugation. The system has since been modified. The reagents are now added automatically and the centrifugation step has been eliminated.

The cost per test for reagents is approximately $1.00. The cost of the luminometer and dispenser/diluter is $6500. When reagent cost, technical time, and cost for plating positive specimens are considered, the cost per test is $2.50.

The Turner® ATP bacterial analysis kit and photometric system (Turner Designs, Mountain View, Calif.) also measures ATP in urine specimens. The Turner® luminometer is designed to measure bacterial ATP by detecting light resulting from bioluminescent oxidation of luciferin in the presence of ATP and luciferase. In this bacteriuria screening system, 0.1 mℓ of urine containing an unknown mixture of microorganisms and somatic cells is placed into 0.9 mℓ of thioglycolate broth. A 25-$\mu\ell$ sample of diluted urine is then placed in 2 tubes. One sample is run immediately (time zero = T[0]) or placed at 0°C for 2 hr and the second is incubated at 37°C for 1 hr. A known ATP standard is run with the test. If the T[0] test sample value is <3% of the ATP standard, the urine is negative and no further testing is required (manufacturer's instructions).

If the T[0] is >30% of the standard, the test is considered positive and the incubated sample need not be run. The incubated sample must be run after 1 hr if the reading at T[0] falls between 3 and 30% of the standard. If the reading of the incubated sample is ≤10% of the standard, the test is interpreted as negative and >10% of the ATP standard or if there is a 2% increase in ATP concentration compared with T[0] reading.

The two previously described bioluminescence methods measure only bacterial ATP by removing somatic ATP. This system is a direct bioluminescence method which combines ATP measurement with growth detection. This does result in a higher false-positive rate since both bacterial and somatic ATP are measured. Since somatic ATP is also measured, this test may also detect pyuria. Further testing is needed to confirm this.

The Turner® bioluminescence method has been compared to the Gram stain, a chemical strip test, a culture method, and a growth-detection automated method.[25] Performance of the direct bioluminescence test compared favorably. It was as sensitive as the automated method and more sensitive than the other two methods at $\geq 10^5$ CFU/mℓ and the Gram stain. Although some specimens require a 1 hr incubation, a majority (62%) tested could be interpreted in less than 1 min after the initial reading.[25] The sensitivity and negative predictive value at 10^4 and 10^5 CFU/mℓ are comparable to other rapid screens (Table 1).

The cost per test of reagents and supplies is approximately $0.50. Since some specimens require repeat testing after 1 hr of incubation, the total cost for those specimens is $1.00. Overall, the average direct cost per specimen, including technical time and repeat testing, is $0.90.[25] If the specimen requires a culture, the cost would be approximately $2.30, which is slightly less than the Lumac® and Monolight systems.

Bioluminescence assays have the advantage of providing rapid results. Detection times vary from 1 min to 1 hr depending on the luminometer used for the assay. Manufacturers recommend that the assay be performed within 6 hr of collection, and if there is a delay of over 30 min, the samples must be stored at 4°C. A small number of specimens were retested by the Lumac® and Microscreen systems after overnight storage at 4°C and the differences were not sufficient to alter the interpretation.[17] However, further testing of a larger specimen population is needed before delayed testing can be recommended.

The cost of instrumentation for the 3 bioluminescence systems described ranges from about $6500 to $8000 (Table 1). The cost for reagents and supplies is comparable and the time for detection is similar. One difference is that the Lumac® and Monolight systems measure bacterial ATP alone while the Turner® system measures both bacterial and somatic ATP, therefore allowing detection of both bacteriuria and pyuria.

A reported disadvantage when using the bioluminescence instruments for urine screening is the increased number of false-positive findings compared to other screening methods. Most reported studies use between 10^4 and 10^5 CFU/mℓ as the definition of positive.[17-21] If bacteria are present at lower levels, the specimen should contain bacterial ATP. Although the number of bacteria present may not be considered clinically significant, the assay will measure the bacterial ATP present in the specimen regardless of the colony count. It has been noted that in actual urine specimens, the average ATP per CFU is considerably higher in specimens with low CFU levels than in those with heavy bacterial growth.[14] False-positive findings may also result from the inability of the culture method to detect growth because of unusual nutrient or anaerobic requirements[14] or when large numbers of leukocytes are present and the nonbacterial ATP cannot be inactivated.[17] The presence of antimicrobial agents may inhibit the growth of bacteria but not inhibit detection by bioluminescence.

Higher false-positive findings occur when the bioluminescence assay is compared to a semiquantitative plate method than when compared to a quantitative pour plate method.[23] A review of earlier studies demonstrate poor correlation with culture results, possibly owing to inconsistent reagent quality,[17] the procedure used to perform the assay, or the reference method used in the study.[23] A comparison of the static manual method and the continuous-flow automated system demonstrate wide variability in the ATP concentration assayed.[23] False-negative findings with the bioluminescence assay result from low levels of bacterial ATP due to low metallic activity of the cells or possible damage to the bacterial cells by antimicrobials or antibody-coated bacteria. Table 1 summarizes the sensitivity, negative predictive value, cost, and detection times of the rapid automated bacteriuria screens.

IV. GROWTH DEPENDENT SYSTEMS

Automated photometry has been described as a method for detecting bacteriuria.[26] This differs from the previously described colorimetric filtration and bioluminescence methods in that growth is required prior to detection and, as a result, detection times are somewhat delayed (30 min to 13 hr). Urine is diluted in a broth medium and incubated at 37°C allowing any bacteria present to grow in the medium. The medium becomes turbid as a result of bacterial growth. The detection of growth is based on changes in light transmission. The change in turbidity is then measured photometrically. Growth occurs within 2 to 3 hr since most organisms causing urinary tract infections are aerobic gram-negative rods. Agitation of the diluted urine will increase growth allowing turbidity to occur more rapidly. Three commercial systems are available for measuring turbidity automatically and therefore have been used as bacteriuria screens.

The Autobac® System (Organon Teknika, Charleston, S.C.), is a semiautomated system which has been used to detect microbial growth.[27-35] Although the system was designed as a rapid method (3 to 5 hr) for antimicrobial susceptibility testing, its use as a rapid urine screen was first described in 1976. The system consists of a photometer for measuring change in turbidity and an incubator shaker. Urine and eugonic broth are inoculated into a multichambered optical cuvette and placed into the incubator shaker. As many as 12 to 18 urine specimens may be screened at once depending upon the number of chambers in the cuvette. The voltage is measured hourly for a maximum of 6 hr. As a result of bacterial growth, the turbidity of the broth increases, causing more light to scatter and the voltage drops. A voltage change of $\geq$0.2 units within 6 hr is considered positive.

The Autobac® MTS system has a sensitivity of at least 90% at 10^4 CFU/mℓ.[29-31,34] The sensitivity of the method is dependent upon incubation time. Studies have shown that the sensitivity increases with longer incubation time.[28-34] A majority (60%) of specimens are positive within 2 hr; however, a maximum incubation of 6 hr is required for the most reliable results.[28] At 4 hr, the sensitivity is approximately 60% at 10^4 CFU/mℓ,[30] although other data show an increase to 91% for the same incubation time.[34] Yeast, especially *Candida albicans,* are not easily detected by the system,[6,27,29,33] and isolates of *Pseudomonas* species require maximum incubation for detection.[29,32,33] Patients on antimicrobial therapy also contribute to the false-negative results,[28,31] along with the use of a liquid boric acid-glycerol-sodium formate preservative for collection of urine specimens.[35]

Cost of screening by the Autobac® system has been evaluated.[31] When compared to other growth detection methods, it was the least expensive screening method. The cost per screen was estimated to be $0.53. This was based on screening ten urines in one cuvette. However, if fewer are screened, the cost would increase.

The AutoMicrobic® system (AMS®, Vitek Systems, Inc., Hazelwood, Mo.) has been designed for both detection and identification, using a miniaturized plastic specimen-handling system, solid-state optics, and a microcomputer. The system consists of a filler-sealer module, reader-incubator module, computer module, and data terminal. For urine screening/identification, the Urine Identification-3 card is used to simultaneously test up to three separate urine samples. A card for screening one urine sample may also be used.

The 91 × 57 mm plastic urine card consists of 30 growth chambers each containing a dehydrated growth medium sealed with oxygen-permeable tape. The wells are divided into three sections and each is rehydrated with diluted urine through a filling port on the side of each section. When reconstituted with diluted urine, each medium will yield a reaction which causes the electro-optics in the reader-incubator to sense the presence or absence of a clinically significant number of organisms.

The media employed in the card are selective for identifying nine commonly isolated urinary pathogens which include *Escherichia coli, Proteus* spp., *Klebsiella/Enterobacter* spp., *Citrobacter freundii, Pseudomonas aeruginosa, Serratia* spp., group D enterococci, *Staphylococcus* spp., and yeast. The manufacturer recommends processing only clean-voided specimens by this method since catheter or suprapubic specimens may contain significant growth at levels below the sensitivity of this system.

The sensitivity for urine screening is $\geq 5 \times 10^4$ CFU/mℓ. The total colony count is reported as $>5 \times 10^4$ CFU/mℓ, 10^3 to 5×10^4 CFU/mℓ, or no growth. Slow-growing organisms, the presence of antimicrobials or transport media containing boric acid may result in false-negative findings. Also, some colony counts between 10^3 and 5×10^4 CFU/mℓ may be reported as "no growth" by the instrument. These limitations should be considered when using the AMS® for urine screening/identification.

Many evaluations have found a high degree of correlation with conventional methods.[31,36-43] The AMS® is the most automated of all systems designed for the microbiology laboratory. The system was first described for urine screening and identification in 1977, although it may be used for other applications. At that time, the sensitivity for both seeded and clinical specimens was reported to be 90%. This was confirmed by a five-laboratory collaborative study.[38] When urine screening was first introduced for this system, only 1 urine specimen could be screened on each card. However, presently 1 to 3 specimens may be tested.

When only 1 urine could be tested by each card, the cost for each negative screen was estimated at $3.23.[31] It had been suggested that the AMS® be used for identification of positive specimens,[40] and another method be used for screening. Use of the Autobac® MTS for screening and the AMS® for identification would be cost-effective[41] when considering growth detection methods. Since three urine specimens may now be simultaneously tested, the cost per test has been reduced to $1.00. This is slightly more expensive than the other growth detection bacteriuria screens,[31] but comparable to the rapid growth-independent systems previously described (Table 1). The AMS® has the advantage of also being an identification system and therefore is the most cost-effective for positive specimens.

Detection is based on the principle of photometry using light emitting diodes and phototransitor detectors to scan the cards at hourly intervals for a maximum of 13 hr. The time required for detection of most urinary pathogens ranges from a mean of 4 hr for *E. coli* to 9 hr for *P. aeruginosa*.[31] A specimen is reported as negative if no change is detected within 13 hr.

The Avantage/MS®-2 (Abbott Laboratories, Diagnostics Division, Irving, Tex.), a growth detection system, may also be used for urine screening. The system consists of an analysis module with light analysis positions and a control module. A successor to the Abbott MS®-2 system is the Avantage which retains the MS®-2 analysis modules and photometric principle of interpretation. The difference between the two instruments is in the computer system which allows improved data handling by the Avantage system and it is slightly easier to operate. Either the MS®-2 system or the Avantage may be used for detection of bacteriuria.

For urine screening, 0.1 mℓ of urine is transferred into an Ampvette (Abbott Laboratories) test unit containing 1 mℓ of eugonic broth. As many as 11 units may be placed into an Ampvette adapter which is inserted into the analysis module allowing the samples to incubate and shake for 5 hr. Growth is measured at 5-min intervals by a series of light emitting diodes and the data are stored in the control module. When a continuous change in light transmission is detected, the specimen is considered positive and the results are printed. Positive results may be detected within 40 min and the test is negative if no turbidimetric change is detected within 5 hr. This urine screening system is similar to the AMS® system in that positive specimens are detected automatically throughout the incubation period.

For pure pathogens at $\geq 10^5$ CFU/mℓ, at least 69% were detected within 2 hr.[31,44] The detection time increased to at least 3 hr for colony counts $\geq 10^4$ CFU/mℓ. The sensitivity of the instrument ranges from 84 to 99% at $\geq 10^5$ CFU/mℓ.[18,19,31,44,45] For those specimens with colony counts between 10^4 and 10^5 CFU/mℓ, the detection rate ranges between 40 to 55%,[31,44,45] and ≤ 31% for lower colony counts.

Table 2
COMPARISON OF GROWTH DEPENDENT SYSTEMS FOR BACTERIURIA

Method	Sensitivity[a] (%) CFU/ml 10^4 All organisms	10^4 Pure pathogens	10^5 All organisms	10^5 Pure pathogens	PVN[a,b] (%) 10^5 CFU/ml	Cost per test[c] ($)	Instrument cost ($)	Detection time (hr)
Autobac	50	90	85	97	99	0.60	30,000[d]	1—6
AMS	51	87	83	96	99	1.00[e]	48,000[f]	4—13
Avantage	45	91	80	96	99	0.60	48,000[g]	1—5

[a] Calculated from reported studies.[27-48]
[b] Predictive value negative.
[c] Screen only.
[d] Photometer and incubator shaker.
[e] Includes identification.
[f] 60 Test module.
[g] Control module and four analysis modules.

Use of a urine collection tube with a liquid boric acid-glycerol-sodium formate preservative may have an effect on the results from the Avantage (MS®-2) in a similar manner as previously described for the Autobac® MTS system. Use of this preservative may result in a delayed detection time and a decreased sensitivity,[46] although one observation did not see a change in the results when urine specimens were collected in containers with and without a preservative.[47] The cost per specimen has been estimated to be less than \$1.00.[19,31] This includes the cost of supplies and technical time. The cost of the instrument varies depending upon the number of analysis modules required for testing within the laboratory.

The *Limulus* amoebocyte lysate (LAL) endotoxin assay has been used with the MS®-2 for urine screening,[48] and found to be an acceptable method for gram-negative bacteriuria at $\geq 10^5$ CFU/ml. The automated assay did not detect gram-positive organisms or yeast and only 40% of the gram-negative organisms between 10^4 and 10^5 CFU/ml. The advantage of using this assay with the Avantage system is that results are available within 20 min as compared to 5 hr with the Avantage alone. The dilution of urine for the study was 1:2000, and it was suggested that a lower dilution might improve the sensitivity of the test for organisms with colony counts below 10^5 CFU/ml. It is not expected that the lower dilution will improve recovery of gram-positive organisms since this method assays endotoxin.

The three growth dependent methods are similar regarding sensitivity, negative predictive value, detection time, and cost (Table 2). The AMS® differs from the Autobac® and Avantage systems in that it also identifies significant urinary pathogens. This requires that incubation time be extended for 13 hr to allow for positive specimens to be detected and identified. Detection occurs within 5 to 6 hr for the other two systems; however, positive cultures require another system for identification and colony counts. The sensitivities of the three systems range from 80% for all organisms at $\geq 10^5$ CFU/ml to >95% for pure pathogens. For colony counts between 10^4 and 10^5 CFU/ml, the sensitivity for all organisms is decreased to 50%. However, this is increased to approximately 90% when only significant urinary pathogens are considered. The negative predictive value for these instruments is 99% at $\geq 10^5$ CFU/ml which indicate that these and other screening methods previously discussed excel in their ability to detect a negative urine. Further studies are needed to determine predictive values for colony counts $<10^4$ CFU/mℓ.

Results from the growth detection systems may be altered by slow-growing organisms, the presence of antimicrobials, transport media containing a boric acid preservative, and organisms with low colony counts. These conditions may produce false-negative results. The false-negative rate for each of these systems is approximately 5% when pure cultures of contaminants or mixed cultures of 2 or more species are excluded from the data analyses. This percentage increases for colony counts below 10^5 CFU/mℓ.

Cost-effectiveness is an important factor, especially when considering instrumentation. The initial cost of the growth detection instruments is much higher than colorimetric filtration and bioluminescence instruments. However, the growth dependent systems may also be used for identification and antimicrobial susceptibility testing. These systems do compare to other automated screening methods when evaluating the cost per test based upon supplies and technical time.

Although the cost per test is comparable to other automated methods for bacteriuria screening, results are somewhat delayed. The need for incubation may limit their use in some laboratories. Positive results may be available within 1 hr, but negative results require 5 to 13 hr. This is more rapid than the overnight plate culture method, but is a delay compared to 2 min for the Bac-T-Screen® and 15 to 30 min for bioluminescence.

Rapid results are meaningful only if they can be reported to the physician in a timely manner. The growth dependent systems provide this mechanism because of their computer capabilities. Screening for bacteriuria by these systems provides an advantage to those laboratories using the instrumentation for other applications.

V. AUTOMATED URINE SCREENING — INVESTIGATIONAL APPLICATIONS

A variety of instrumentation for detection of bacteriuria has been described, but has not been applied to routine screening of urine specimens in the diagnostic microbiology laboratory. These chemical and physical methods include electrochemical,[49] electrical impedance and conductivity,[50-55] flow cytometry,[56] laser nephelometry,[57] microcalorimetry,[58] particle size distribution analysis,[59-62] and radiometry.[63]

The electrochemical method is based on measuring a change in potential between two electrodes. A reference electrode and a platinum electrode connected to a strip-chart recorder are needed for performance of the test. These electrodes are placed directly into a broth containing urine, and the presence of viable bacteria are detected by measuring a voltage change. The detection time is defined as the time elapsed between the inoculation of urine into a phenol red broth with glucose and the initial voltage increase detected by the strip-chart recorder.

The detection time for this method ranged from 2 to 9 hr for seeded samples,[49] and from 9 min to 10 hr for patient specimens with $\geq 10^5$ CFU/mℓ. A majority (94%) of positive specimens were detected within 4 hr. It has been suggested[49] that some of the false-negative results may be due to inhibitory factors present in urine such as antimicrobics. Although this has been suggested as a screening method for clinics when complete microbiology facilities are not available, the time required for detection may not be impractical in those situations. Further studies are needed on larger patient populations to evaluate the sensitivity of this method at low colony counts and to assess the cost for screening.

Electrical impedance has also been applied to bacteriuria screening.[50-54] This method measures resistance to the electrical flow of current passed through a solution due to bacterial growth and metabolism. Initially, the method required filtering urine to trap bacteria,[50] then measuring impedance by using a four-electrode system. This was done to remove leukocytes and other debris to avoid interference in the system. Detection of bacteriuria by this method had a 90% correlation with the culture method at $\geq 10^5$ CFU/m*l*. This system also detected between 10^3 and 10^4 CFU/m*l*. This is a static assay and does not depend on metabolic activity of organisms.

Filtration of urine has been eliminated when using the Bactometer® microbial monitoring system (Bactomatic, Inc., Princeton, N.J.). The system is composed of two identical chambers, a reference and test chamber containing an identical medium.[51] The test chamber is inoculated with 0.5 m*l* of urine and both chambers are incubated at 35°C in the instrument. Using this method,[52] there was >96% agreement with colony counts $\geq 10^5$ CFU/m*l* within 4 hr and 94% correlation at 2 hr. As microorganisms grow and metabolize, they change the composition of the medium. This results in a change in the resistance to the flow of an alternating current when a pair of electrodes are placed into the medium.[53,54] Using 2.6 hr as a cutoff, the correlation with culture was 95.8%.[54] The false-negative rate was 14%, which may be due to the presence of antimicrobials or slow-growing organisms. When the cutoff was increased to 3.5 hr, the false-negative rate decreased, but, as expected, the number of false-positive specimens increased.

A more recent development which monitors conductance rather than impedance is the *Malthus* system (Malthus Instruments Ltd., Perthshire, Scotland). This system uses ceramic electrodes to measure conductance in a 1:10 dilution of urine and brain heart infusion broth. An accelerating change in conductance of 10 μS was considered positive. When this system was compared to the Bactometer®,[55] the least number of false-positives and false-negatives occurred at a cutoff of 4 hr. Both systems are simple to operate,[55] and microcomputers are available for automatic readings. The cost of supplies for the Bactometer® is approximately $0.55 compared to $0.30 for the Malthus system. The difference is that Bactometer® modules are disposable, while Malthus modules are reusable. The approximate cost of a Bactometer® instrument and chart recorder is $14,000. These systems seem to have the same limitations as the growth dependent methods previously described. These include a delayed detection time of at least 4 hr, a high false-negative rate below 10^5 CFU/m*l*, and poor detection of slow-growing organisms. However, they do have the ability to detect negative specimens with confidence.

Flow cytometry has been described for detecting bacteria in urine.[56] A double fluorescent DNA staining method is applied to the bacteria and fluorescence from the stained organisms is measured with a dual-beam flow cytometer. The bacteria stain, and there is minimal nonspecific staining of other material in the urine specimen. More work is needed to determine the predictive values of this method as well as its practical application in a diagnostic microbiology laboratory.

Laser nephelometry is similar to the principles of photometry except that a heme laser is used as the light source. Light is scattered by particles or bacteria growing in a broth medium. The scattered light is focused by a lens system onto a photodiode and an electrical signal is displayed. This results in a voltage increase which is then measured.

For urine screening,[57] 0.1 m*l* of urine is inoculated into 1 m*l* of eugonic broth and allowed to incubate. Voltage readings are taken at timed intervals and a specimen is considered positive when there is a voltage increase of 0.2 V. When compared to a semiquantitative plate culture method,[57] the sensitivity of laser nephelometry at 10^5 CFU/m*l* was 96% which decreased to 75% for specimens containing between 10^4 and 10^5 CFU/m*l*. The negative predictive value for specimens with at least 10^4 CFU/m*l* was 94%. The mean detection time for commonly isolated urinary pathogens at 10^5 CFU/mℓ is slightly longer than 1 hr. This is comparable to the Autobac® and Avantage systems. A specimen is considered negative if there is not significant increase in voltage within 6 hr. This delayed detection time also compares to other growth detection methods.

Microcalorimetry[58] has been applied to bacteriuria screening. This method measures the heat generated by microbial metabolism. When compared to electrochemical, electrical impedance, and bioluminescence methods,[54] it had the highest number of false-positive results, but the least number of false-negatives. The method is dependent on growth and metabolism of organisms and therefore results are somewhat delayed.

Rapid screening of urine by particle size distribution analysis[59-62] has been described. Initial work[60] showed that this method lacked sensitivity because of many interfering substances which may be present in urine and, therefore, would be counted. The sensitivity was increased by adding volume distribution analysis,[61] however this method was not practical for routine use in a diagnostic laboratory. These particles can be dispersed by placing the urine in a sonicating water bath or use of a sonic probe prior to testing[59] in order to increase the sensitivity. A counting time of 15 sec may be used. False-positive results may be due to large numbers of leukocytes in the specimen; however, this may be a significant finding in the diagnosis of a urinary tract infection.

Instrumentation which may be used for this analysis include a Coulter® Counter (Model ZBI) linked to a C1000 Channelyzer (Coulter Electronics).[62] The sensitivity of this method was high when the colony count was at least 10^8 CFU/mℓ. As the number of organisms in the specimen decreases, the sensitivity also greatly decreases. This method has the potential of being a rapid (5 to 10 min) urine screen if the sensitivity could be improved. A suggested modification[62] would be to prefilter the urine specimen. Although this would add a delay and increase the cost, it may be a necessary step to improve the method for use in a diagnostic laboratory.

The use of radiometry[63] for the detection of bacteriuria has been under investigation. The Bactec® system (Johnston Laboratories, Becton Dickinson & Co., Cockeysville, Md.) measures the release of $^{14}CO_2$ from radiolabeled ^{14}C substrate that occurs as a result of bacterial metabolism. A sample of urine is placed in a broth containing ^{14}C and allowed to incubate and shake for 1 to 2 hr. A reading of 30 growth index units or greater is considered positive. More work is needed to determine its accuracy and cost-effectiveness, especially for low-level bacteriuria.

A review of automated methods for detection of bacteriuria demonstrates a number of rapid, accurate, and cost-effective systems. The gold standard regarding the definition of a urinary tract infection has been $\geq 10^5$ CFU/mℓ. This definition seems acceptable when screening the asymptomatic patient; however, further considerations at $\geq 10^5$ CFU/mℓ and pyuria is necessary for the symptomatic patient. Presently, the sensitivity at these low levels are not known for most of the automated urine screens. Future investigations should include these data so that these methods may be used with confidence to screen all patients. Until these data become available, careful consideration should be given to the patient population before a screening method is selected. Knowledge of the limitation of these methods will allow for more complete testing by other methods.

REFERENCES

1. Kass, E. H., Asymptomatic infections of the urinary tract, *Trans. Assoc. Am. Physicians,* 69, 56, 1956.
2. Bartlett, R. C. and Galen, R. S., Predictive value of a urine culture, *Am. J. Clin. Pathol.,* 79, 756, 1983.
3. Stamm, W. E., Counts, G. W., Running, K. R., Fihn, S., Turck, M., and Holmes, K. K., Diagnosis of coliform infection in acutely dysuric women, *N. Engl. J. Med.,* 307, 463, 1982.
4. Stamm, W. E., Wagner, K. F., Amsel, R., Alexander, E. R., Turck, M., Counts, G. W., and Holmes, K. K., Causes of acute urethral syndrome in women, *N. Engl. J. Med.,* 303, 409, 1980.
5. Davis, J. R., Stager, C. E., and Aroj, G. F., Clinical laboratory evaluation of a bacteriuria detection device for urine screening, *Am. J. Clin. Pathol.,* 81, 48, 1984.
6. Pezzlo, M. T., Wetkowski, M. A., Peterson, E. M., and de la Maza, L. M., Evaluation of a two-minute test for urine screening, *J. Clin. Microbiol.,* 18, 697, 1983.
7. Pezzlo, M. T., Detection of bacteriuria by automated methods, *Lab. Med.,* 15, 539, 1984.
8. Cashman, J. S., Ahlin, P. A., and Matsen, J. M., Evaluation of a rapid bacteriuria screening instrument, *Diagn. Microbiol. Infect. Dis.,* 2, 11, 1984.
9. Pfaller, M. A., Baum, C. A., Niles, A. C., and Murray, P. R., Clinical laboratory evaluation of a urine screening device, *J. Clin. Microbiol.,* 18, 674, 1983.
10. Wallis, C., Melnick, N. L., and Longoria, C. J., Colorimetric method for rapid determination of bacteriuria, *J. Clin. Microbiol.,* 14, 342, 1981.
11. Pezzlo, M. T., Wetkowski, M. A., Peterson, E. M., and de la Maza, L. M., Detection of bacteriuria and pyuria within two minutes, *J. Clin. Microbiol.,* in press.
12. Daifukus, R., Zeleznik, D., and Stamm, W. E., Detection of bacteriuria and pyuria with the Bac-T-Screen bacteriuria and pyuria with the Bac-T-Screen bacteriuria detection device, Paper 1072, *Abstracts of 1984 ICAAC Meeting,* American Society for Microbiology, Washington, D.C., 1984.
13. McElroy, W. D. and Ballentine, R., The mechanism of bioluminescence, *Proc. Natl. Acad. Sci. U.S.A.,* 30, 377, 1944.
14. Thore, A. S., Ånséhn, S., Lundin, A., and Bergman, S., Detection of bacteriuria by luciferase assay of adenosine triphosphate, *J. Clin. Microbiol.,* 1, 1, 1975.
15. Alexander, D. N., Ederer, G. M., and Matsen, J. M., Evaluation of adenosine 5′-triphosphate assay as a screening method to detect significant bacteriuria, *J. Clin. Microbiol.,* 3, 42, 1976.
16. Thore, A., Lundin, A., and Ånséhn, S., Firefly luciferase ATP assay as a screening method for bacteriuria, *J. Clin. Microbiol.,* 17, 218, 1983.
17. Drow, D. L., Baum, C. H., and Hirschfield, G., Comparison of the Lumac and Monolight systems for detection of bacteriuria by bioluminescence, *J. Clin. Microbiol.,* 20, 797, 1984.
18. Szilagyi, G., Aning, V., and Karmen, A., Comparative study of two methods for rapid detection of clinically significant bacteriuria, *J. Clin. Lab. Automat.,* 3, 117, 1983.
19. Welch, W. D., Layman, M. A., and Southern, P. M., Evaluation of the MS-2 and Lumac systems for the rapid screening of urine specimens, *Am. J. Clin. Pathol.,* 81, 629, 1984.
20. Park, C. H., Hixon, D. L., McClintock, C. C., Ferguson, C. B., Lawless, C. C., Rishem, C. C., and Cook, C. B., Rapid detection of insignificant bacteriuria by concomitant use of Lumac system and Gram's stain, *Am. J. Clin. Pathol.,* 82, 593, 1984.
21. Welch, W. D., Thompson, L., Layman, M., and Southern, P. M., Evaluation of two bioluminescence-measuring instruments, the Turner Design instrument and the Lumac system, for rapid screening of urine specimens, *J. Clin. Microbiol.,* 20, 1165, 1984.
22. Mackett, D., Kessock-Philip, S., Bascomb, S., and Easomon, C. S. F., Evaluation of the Lumac kit for the detection of bacteriuria by bioluminescence, *J. Clin. Pathol.,* 35, 107, 1982.
23. Nichols, W. W., Curtis, G. D., and Johnston, H. H., Analysis of the disagreement between automated bioluminescence-based and culture methods for detecting significant bacteriuria, with proposals for standardizing evaluations of bacteriuria detection methods, *J. Clin. Microbiol.,* 15, 802, 1982.
24. Hanna, B. A., unpublished data, 1984.
25. Schifman, R. B., Wieden, M., Brooker, J., Cherry, M., Delduca, M., Norgard, K., Palen, C., Reis, N., Swanson, J., and White, J., Bacteriuria screening by direct bioluminescence assay of ATP, *J. Clin. Microbiol.,* 20, 644, 1984.
26. O'Grady, F., Pearson, N. J., Ridout, P., Murphy, B., MacDonald, T., and Johnson, F., Automated photometry for detecting bacteriuria, in *Rapid Methods and Automation in Microbiology,* Tilton, R. C., Ed., American Society for Microbiology, Washington, D.C., 1981, 166.
27. Thrupp, L. D., Heinze, P. D., and Pezzlo, M. T., Direct rapid urine cultures: same day reporting of quantitation, speciation and antimicrobial susceptibility tests, in *ASM Symp. Automation in Clinical Microbiology,* Balows, A., Ed., Pfizer Diagnostics, New York, 1976, 17.

28. Jenkins, R. D., Hale, D. C., and Matsen, J. M., Rapid semiautomated screening and processing of urine specimens, *J. Clin. Microbiol.*, 11, 220, 1980.
29. Hale, D. C., Wright, D. N., McKie, J. E., Isenberg, H. D., Jenkins, R. D., and Matsen, J. M., Rapid screening for bacteriuria by light scatter photometry (Autobac): a collaborative study, *J. Clin. Microbiol.*, 13, 147, 1981.
30. Kelly, M. T. and Balfour, L. C., Evaluation and optimization of urine screening by Autobac, *J. Clin. Microbiol.*, 13, 677, 1981.
31. Pezzlo, M. T., Tan, G. L., Peterson, E. M., and de la Maza, L. M., Screening of urine cultures by three automated systems, *J. Clin. Microbiol.*, 15, 468, 1982.
32. Heinz, T., Pezzlo, M., and Thrupp, L., Clinical and laboratory efficiency of rapid screening culture by nephelometry, *Am. J. Clin. Pathol.*, 77, 305, 1982.
33. Hale, D. C., Thrupp, L. D., and Matsen, J. M., Evaluation of urine culture screening by light scatter photometry, *Am. J. Clin. Pathol.*, 76, 208, 1981.
34. Crout, F. V. and Tilton, R. C., Rapid screening of urine for significant bacteriuria by Gram stain, acridine orange stain, and the Autobac MTS system, *Diagn. Microbiol. Infect. Dis.*, 2, 179, 1984.
35. Pezzlo, M. T., Conway, S., Jacobson, M., Peterson, E. M., and de la Maza, L. M., Effect of the B-D urine culture kit on an automated bacteriuria screen, *J. Clin. Microbiol.*, 20, 1207, 1984.
36. Pezzlo, M. T., Automated methods for detection of bacteriuria, *Am. J. Med.*, July 28(Suppl), 76, 1983.
37. Aldridge, C., Jones, P. W., Gibson, S., Lanham, J. L., Meyer, M., Vannest, R., and Charles, R., Automated microbiological detection/identification system, *J. Clin. Microbiol.*, 6, 406, 1977.
38. Isenberg, H. D., Gavan, T. L., Sonnenwirth, A., Taylor, W. I., and Washington, J. A., II, Clinical laboratory evaluation of automated microbial detection/identification system in the analysis of clinical urine specimens, *J. Clin. Microbiol.*, 10, 226, 1979.
39. Nicholson, D. P. and Koepke, J. A., The AutoMicrobic system for urine, *J. Clin. Microbiol.*, 10, 823, 1979.
40. Hamoudi, A. C. and Lin, S., Cost analysis of the AutoMicrobic System urine identification card, *J. Clin. Microbiol.*, 14, 411, 1981.
41. Wadke, M., McDonnell, C., and Ashton, J. K., Rapid processing of urine specimens by urine screening and the AutoMicrobic system, *J. Clin. Microbiol.*, 16, 668, 1982.
42. Smith, P. B., Gavan, T. L., Isenberg, H. D., Sonnenwirth, A., Taylor, W. I., Washington, J. A., II, and Balows, A., Multi-laboratory evaluation of an automated microbial detection/identification system, *J. Clin. Microbiol.*, 8, 657, 1978.
43. Sonnenwirth, A. C., Preprototype of an automated microbial detection and identification: a developmental investigation, *J. Clin. Microbiol.*, 6, 400, 1977.
44. McCarthy, L. R., Gavan, T. L., Robson, J., and Corlett, C., Evaluation of the MS-2 urine screening method for detection of bacteriuria, *J. Clin. Microbiol.*, 16, 250, 1982.
45. Hoban, D. J., Koss, J. C., Gratton, C. A., and Ronald, A. R., Urine screening with the MS-2, *J. Clin. Microbiol.*, 17, 1061, 1983.
46. Hubbard, W. A., Shalis, P. J., and McClatchey, K. D., Comparison of the B-D urine culture kit with a standard method and with the MS-2, *J. Clin. Microbiol.*, 17, 327, 1983.
47. Southern, P. M. and Luttrell, B., Use of the Becton-Dickinson urine culture tube with the Abbott MS-2 urine screening system, *Diagn. Microbiol. Infect. Dis.*, 2, 193, 1984.
48. Jorgensen, J. H. and Alexander, G. A., Rapid detection of significant bacteriuria by use of an automated *Limulus* amoebocyte lysate assay, *J. Clin. Microbiol.*, 16, 587, 1982.
49. Lamb, V. A., Dalton, H. P., and Wilkins, J. R., Electrochemical method for the early detection of urinary tract infections, *Am. J. Clin. Pathol.*, 66, 91, 1976.
50. Wheeler, T. G. and Goldschmidt, M. C., Determination of bacterial cell concentrations by electrical measurements, *J. Clin. Microbiol.*, 1, 25, 1975.
51. Hadley, W. K. and Senyk, G., Early detection of microbial metabolism and growth by measurement of electrical impedance, in *Microbiology 1975*, Schlessinger, D., Ed., American Society for Microbiology, Washington, D.C., 1975, 12.
52. Zafari, Y. and Martin, W. J., Comparison of the Bactometer microbial monitoring system with conventional methods for detection of microorganisms in urine specimens, *J. Clin. Microbiol.*, 5, 545, 1977.
53. Throm, R., Specter, S., Strauss, R., and Friedman, H., Detection of bacteriuria by automated electrical impedance monitoring in a clinical microbiology laboratory, *J. Clin. Microbiol.*, 6, 271, 1977.
54. Cady, P., Dufour, S. W., Lawless, P., Nunke, B., and Kraeger, S. J., Impedimetric screening for bacteriuria, *J. Clin. Microbiol.*, 7, 273, 1978.
55. Brown, D. F. J. and Warner, M., Impedance and conductivity methods for detecting bacteriuria, in *Rapid Methods and Automation in Microbiology*, Tilton, R. C., Ed., American Society for Microbiology, Washington, D.C., 1981, 171.

56. Van Dilla, M. A., Langlois, R. G., Pinkel, D., Yajko, D., and Hadley, W. K., Bacterial characterization by flow cytometry, *Science*, 220, 620, 1983.
57. Richet, H., Bayardelle, P., and Lemarie, J. C., Rapid detection of bacteriuria by laser nephelometry, *J. Clin. Microbiol.*, 19, 240, 1984.
58. Beezer, A. E., Bettelheim, E. D., Newell, E. D., and Stevens, J., The diagnosis of bacteriuria by flow microcolorimetry, a preliminary investigation, *Sci. Tools*, 21, 13, 1974.
59. Alexander, M. K., Khan, M. S., and Dow, C. S., Rapid screening of urine for infection by particle size distribution analysis, in *Rapid Methods and Automation in Microbiology*, Tilton, R. C., Ed., American Society for Microbiology, Washington, D.C., 1981, 176.
60. Smither, R., Rapid screening for significant bacteriuria using a Coulter Counter, *J. Clin. Pathol.*, 30, 1158, 1977.
61. Dow, C. S., France, A., Khan, M., and Johnson, T., Particle size distribution analysis for the rapid detection of microbial infection of urine, *J. Clin. Pathol.*, 32, 386, 1979.
62. Baker, J. S., Ederer, G. M., and Mundschenk, D., Particle size distribution analysis as a rapid method to detect significant bacteriuria, *Am. J. Med. Technol.*, 49, 727, 1983.
63. Leader, I. and Potts, R., Rapid diagnosis of urinary tract infections by Bactec and Gram stain, Abstr. 3rd Int. Symp. Rapid Methods and Automation in Microbiology, Washington, D.C., Session 4, paper 11, 1981.

Chapter 4

RADIOMETRIC DETECTION, IDENTIFICATION, AND ANTIMICROBIAL SUSCEPTIBILITY TESTING OF MYCOBACTERIA

Margie A. Morgan and Glenn D. Roberts

TABLE OF CONTENTS

I. INTRODUCTION

The growth, identification, and susceptibility testing of mycobacteria using conventional methods are time consuming, primarily due to the characteristic slow growth of many members of the genus. In an era when "clinically significant" reports and decreased time to reporting are important considerations, an instrument which has the capability of decreasing the time to growth detection and identification is an important advantage. This instrument is further appreciated if the decreased reporting time is coupled with comparable or decreased cost of test performance compared to conventional methods. This chapter is devoted to the topic of development of automation in the mycobacteriology laboratory with the Bactec® system (Johnston Laboratories, Towson, Md.). Included is a discussion of the development, operation, and evaluation of the Bactec® system for mycobacterial growth detection, identification, and antimicrobial susceptibility testing of *Mycobacterium tuberculosis* and our opinion of this system for routine use in the clinical mycobacteriology laboratory.

II. THE DEVELOPMENT OF THE BACTEC® SYSTEM

Radiometric methods were first introduced in microbiology in 1960 by Deland and Wagner,[1] who developed a technique for automated detection of microbial growth and metabolism by measuring the conversion of ^{14}C-labeled substrate (primarily ^{14}C-glucose) to $^{14}CO_2$. A radiometric system capable of detecting bacterial growth in blood cultures was subsequently made available to the clinical laboratory with the Bactec® 250. Technical advances over the last 10 years resulted in development of the model presently in use, the Bactec® 460 (Figure 1). Cummings et al.[2] in 1975 modified the principle of radiometric detection of microbial growth to the detection of the metabolism of *M. tuberculosis* in pure culture. The modification changed the substrate from ^{14}C-glucose to ^{14}C-glycerol or ^{14}C-acetate, since they are more readily metabolized by mycobacteria. Middlebrook 7H10 medium (agar free, glycerol free) was used for organism recovery.

The development of a selective broth medium for the recovery of mycobacteria that could be used with clinical specimens in an automated system greatly contributed to the development of the Bactec® system for mycobacterial detection. The radiometric detection method developed by Cummings et al.[2] used agar-free, glycerol-free Middlebrook 7H10 medium for mycobacterial growth detection, but this nonselective medium was not adequate for clinical specimens. In 1972, Mitchison et al.[3] modified Middlebrook 7H10 oleic acid albumin agar medium by adding antibiotics (polymyxin B, carbenicillin, amphotericin B, and trimethoprim) to make a selective recovery medium. Mitchison et al.[4] used the same antibiotic combination in Middlebrook 7H11 agar base for the recovery of mycobacteria from tissue specimens not treated with the usual alkaline decontaminating agents. A less inhibitory form of Middlebrook 7H11 agar was developed by McClatchey et al.[5] by reducing the amount of carbenicillin. In 1977, Middlebrook[6] developed 7H12 broth medium, an enriched Middlebrook 7H9 base with the addition of casein hydrolysate and the antibiotic combination described by McClatchey. The selectivity of 7H12 broth for mycobacteria was increased by deleting glucose and glycerol from the medium. The selective Middlebrook 7H12 broth was found to be suitable for the automated detection of mycobacteria in clinical specimens and became the primary recovery medium for the Bactec® system. ^{14}C-palmitic acid is the radiolabeled substrate presently used for mycobacterial growth detection.

III. THEORY AND OPERATION OF THE BACTEC® 460

Preparation of clinical specimens for use in the Bactec® method does not signifi-

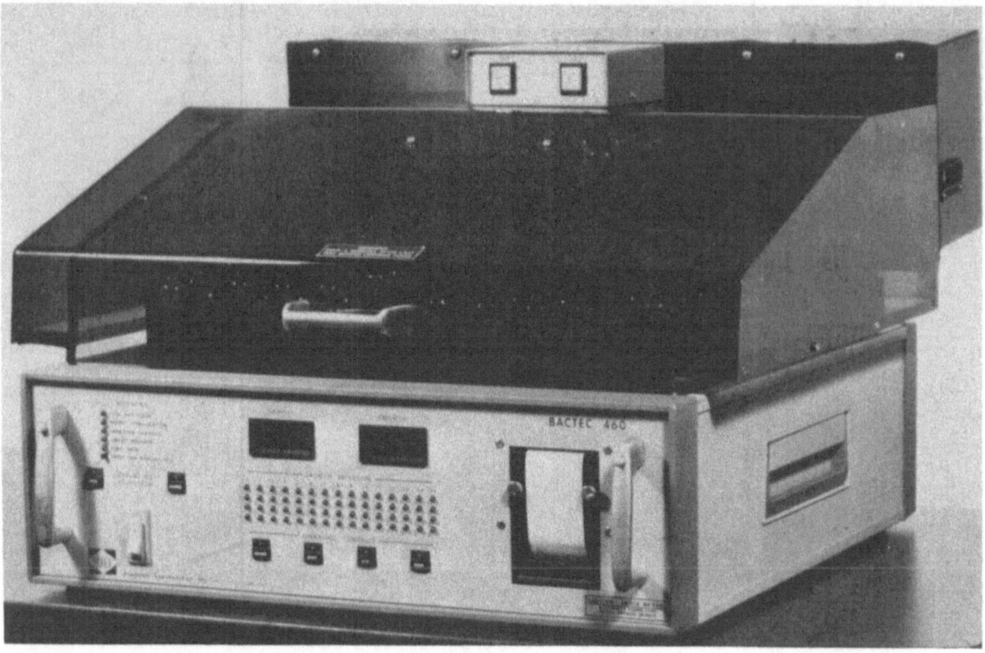

FIGURE 1. Bactec® Model 460 equipped with TB hood.

cantly differ from conventional decontamination and culturing methods. Sterile body fluids and tissues (0.1 mℓ) are inoculated directly into the Bactec® culture vials. Potentially contaminated specimens are first decontaminated using a preferred decontamination method. Sodium hydroxide (2 to 4%) with or without mucolytic agents, trisodium phosphate, or oxalic acid are suitable decontamination methods. Following decontamination and concentration by centrifugation, 0.1 mℓ of specimen is inoculated into a culture vial using a needle and syringe. The vials should be inoculated in a biological safety cabinet to avoid exposure to aerosols. Extreme care should be exercised during the inoculation process with needle and syringe.

The culture vials contain 2 mℓ of Middlebrook 7H12 broth containing ^{14}C-palmitic acid and are sealed with a rubber septum. When introduced into the instrument, two sterile needles puncture the rubber septum and penetrate into the gas space below the septum, but above the medium in the vial (Figure 2). The ion chamber of the radioactive detection device contained in the system is partially evacuated with a pump when an inlet valve is opened, atmospheric gas is drawn from the separate gas cylinder (5% CO_2, 95% compressed air) through a sterilizing filter, into the vial by one needle and out of the vial by the second needle. Therefore, the atmosphere from the culture vial is sampled and passed into the ion chamber. This process evacuates the radioactive atmosphere from the test vial and refills the vial with sterile gas from the cylinder. The ion chamber analyzes the test vial gas, the radioactivity is expressed as an ionization current and is amplified, then digitized. The digital readout is printed on paper tape and is also displayed on the front panel of the instrument. The amount of radioactivity present in the test gas is measured and displayed as index units, where a reading of 100 equals approximately 0.025 μCi ^{14}C activity. A reading of 10 or greater is usually considered positive for mycobacterial growth. When this threshold is reached, the positive samples are identified on the printout and on the front panel of the machine by a red warning light. The needles are withdrawn from the vial following atmosphere sam-

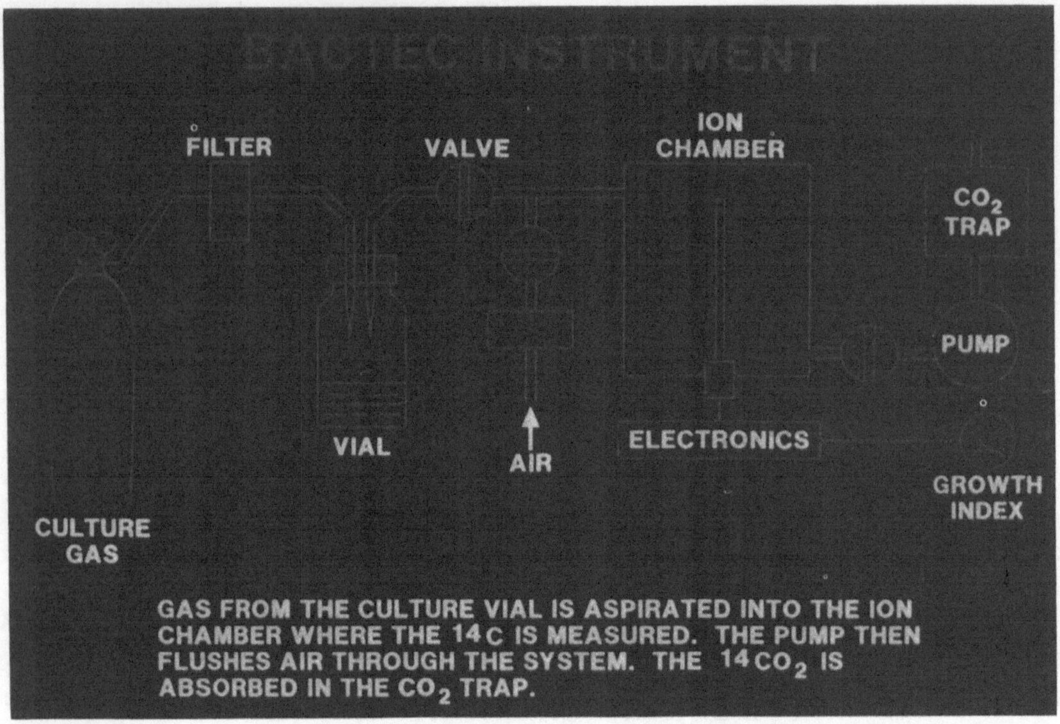

FIGURE 2. Schematic diagram illustrating the principle of radiometric detection of microorganisms.

pling, and the ion chamber is flushed into a carbon dioxide trap to remove any residual radioactivity. To eliminate cross-contamination of vials with microbial organisms, a heating unit sterilizes the needles before moving to the next specimen vial. A self-contained safety hood, purchased separately from the detection unit, is the only modification necessary to use the Bactec® 460 blood culture system for mycobacterial detection.

Before disposal, the vials should be autoclaved at 121°C and 15 lb of pressure for 1 hr. One should avoid excessive moving of the bottles until cooling to decrease the hazard of breakage. Since the vials contain a small amount of radioactivity (2 μCi/vial), each Bactec® user should review their state and institution regulations concerning the disposal of low-level radioactive material.

IV. EVALUATION OF THE BACTEC® SYSTEM

Several studies comparing the Bactec® method to conventional culturing methods have been reported. The evaluations have been divided into two groups based on the acid-fast smear positivity or negativity. Roberts and co-workers[7] reported the Bactec® system to be as successful as conventional culturing methods in detecting growth in specimens having a positive direct smear for acid-fast bacilli. A total of 370 respiratory specimens were detected as culture positive either by conventional culture methods or by the radiometric technique. The radiometric method was more successful in detecting growth of *M. tuberculosis* and mycobacteria other than *M. tuberculosis* (MOTT) than conventional cultural methods (Figure 3). The mean detection times for Bactec® positive cultures for *M. tuberculosis* and MOTT were 8.3 and 5.2 days, respectively. Conventional methodology provided growth detection in 9.4 days for *M. tuberculosis* and 17.8 days for MOTT. The Bactec® system has not been as successful in detecting growth in specimens with a negative acid-fast smear. Morgan et al.[8] reported 71 acid-

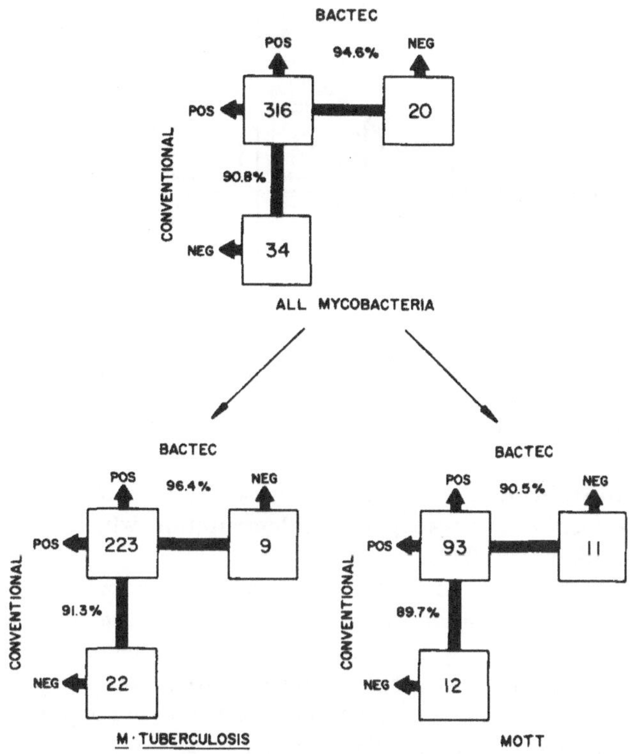

FIGURE 3. Comparison of Bactec® and conventional methods for the recovery of 370 culture-positive specimens. (From Roberts, G. D. et al., *J. Clin. Microbiol.*, 18, 689, 1983. With permission.)

fast culture positive specimens recovered from 2165 submitted acid-fast smear negative specimens. The Bactec® system detected 71.8% of the cultures where conventional cultural procedures detected 88.7% (Figure 4). It was felt that the lower detection rate in this specimen group was due to the amount of specimen cultured. One vial of Middlebrook 7H12 broth medium was inoculated with 0.1 mℓ of inoculum, while a total of 1.5 mℓ was used with the conventional method (two tubes of Lowenstein-Jensen medium, [0.25 mℓ/tube] and one plate each of Middlebrook 7H10 and Middlebrook 7H11 [0.5 mℓ/plate]). By increasing the Middlebrook 7H12 broth medium to 4 mℓ per vial and the inoculum size to 0.5 mℓ, the percentage of mycobacteria detected using 1 vial approached the detection rate of conventional culture methods for both acid-fast smear negative and positive specimens.[9] However, the contamination rate in the 4 mℓ vial was approximately twice that of the 2 mℓ vial with 0.1 mℓ inoculum. If the contamination rate in the 4 mℓ vial could be reduced, a system using this vial would provide greater sensitivity for the detection of acid-fast smear negative specimens.

As of June 1985, an evaluation of a 4 mℓ vial of Middlebrook 7H12 medium containing polymyxin, amphotericin B, nalidixic acid azlocillin, and trimethoprim inoculated with 0.5 mℓ of specimen provided a contamination rate of 2.3% compared to 5.8% and 1.4% using the 2 mℓ vial currently recommended and conventional methods, respectively. Most of the contamination is attributed to *Pseudomonas aeruginosa* growing after 1 to 4 days of inoculation of the clinical specimen. All vials registering a positive growth index must be sampled and stained with an acid-fast stain to confirm the presence of mycobacteria and subcultured to appropriate medium to exclude the presence of bacterial species. At present, bacterial contamination is a significant

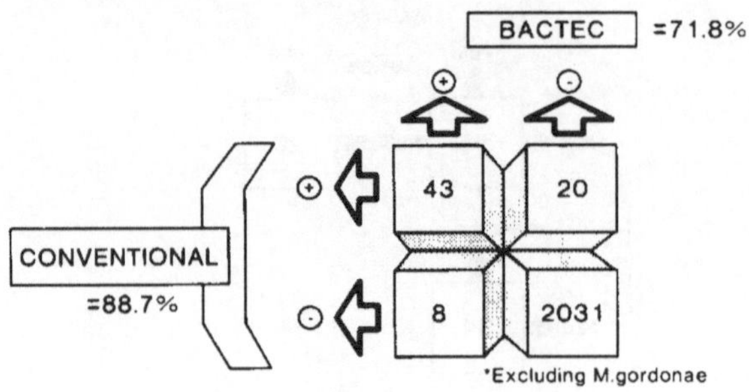

FIGURE 4. Comparison of the number of mycobacteria recovered from 71 acid-fast, smear-negative specimens by the Bactec® and conventional methods. (From Morgan, M. A. et al., *J. Clin. Microbiol.*, 18, 384, 1983. With permission.)

enough problem to warrant the use of the Bactec® 460 concurrently with at least one conventional culture medium. We suggest supplementation with either Middlebrook 7H10 or selective Middlebrook 7H11.

The outstanding feature of the Bactec® system is the decreased time to detection of mycobacterial growth as compared with conventional culture methods. However, when reviewing these evaluations, one must remember that the detection times quoted do not include time required for subculture and identification. The use of the broth medium in this system requires subculturing before colonies are available for observation or biochemical testing. Therefore, the inability to distinguish a clinically significant isolate from a contaminant *(M. gordonae)* at the time of detection is clinically important and presents a practical problem in using the Bactec® system. Without species identification, the information provided at the time of detection is somewhat equal in importance to the clinician as a positive acid-fast smear result, or perhaps less if one considers the high rate of detection of *M. gordonae* by the Bactec® system. In an evaluation at the Mayo Clinic, the recovery of *M. gordonae* approached 60% of total isolates in the Bactec® system.[9] If rapid identification procedures could be developed for the species identification of mycobacteria detected by the Bactec® system, the relevance of the information would be increased and the time-saving potential of the Bactec® could be better appreciated.

V. RADIOMETRIC IDENTIFICATION OF *MYCOBACTERIUM TUBERCULOSIS*

A radiometric identification test for *M. tuberculosis* has been developed which differentiates *M. tuberculosis* complex (*M. tuberculosis, M. bovis,* and *M. africanum*) from other mycobacterial species by the differential growth pattern in the chemical *p*-nitro-α-acetyl-amino-β-hydroxy-propiophenone (NAP).[10] The methodology includes inoculation of the test organism into one control bottle containing Middlebrook 7H12 medium and one test bottle containing 7H12 medium supplemented with NAP. The source for the organism can be either a culture growing on agar medium (indirect method) or a positive Bactec® culture vial proven by acid-fast stain to contain mycobacteria (direct method). The performance of the direct test is dependent upon a standardized inoculum and may require additional days of growth past initial detection by the Bactec® system or dilution of the original vial. Once the appropriate inoculum is obtained, the bottles are assayed on the system for no more than 5 consecutive days, and the growth index is recorded. Mycobacteria in the tuberculous complex are unable to grow in the presence of NAP after 5 days of incubation and will not exhibit any significant increase in the growth index compared to the control bottle. Other species of mycobacteria grow well in the presence of NAP and will show a marked increase in the growth index compared to the control test vial. One possible exception to this procedure is with isolates of *M. kansasii.* Some isolates of *M. kansasii* do not grow well initially in the NAP solution and require 7 days of incubation for interpretable result.[11] Both Hemple and Goodman[12] and Linder and Larsh[13] compared NAP testing with conventional biochemical testing for the identification of the tuberculous complex in clinical isolates and found 100% agreement. Laszlo and Siddiqi[14] studied 416 coded cultures (234 belonging to *M. tuberculosis* complex and 182 belonging to 35 other mycobacterial species) in the indirect and direct NAP test. Their data also showed NAP to be effective in separating the *M. tuberculosis* complex from other mycobacterial species. Our experience has also been favorable; testing 400 clinical isolates in either the indirect or direct method, only one *M. tuberculosis* strain was misidentified as MOTT.[11] One advantage of the NAP procedure is the correct identification of multiply antituberculous-drug-resistant strains of *M. tuberculosis* which may show a negative reaction in the niacin test. The NAP test is helpful clinically because the rapid confirmation of *M. tuberculosis* isolates is important from a public health standpoint and for the initiation of proper therapy for the index case and prophylaxis for individuals having contact with the infected person.

VI. RADIOMETRIC DETERMINATION OF DRUG SUSCEPTIBILITY TESTING FOR MYCOBACTERIA

The Bactec® system provides a rapid method for drug susceptibility testing of *M. tuberculosis.* Test organisms are inoculated into vials containing Middlebrook 7H12 broth and antimicrobials. The inoculum may be taken directly from a Bactec® vial (direct method) or from an actively growing culture on an agar medium (indirect method). Control bottles contain a 1:100 dilution of inoculum and no antimicrobials to make the methodology equivalent to the 1% proportion method commonly used by most laboratories. The inoculated vials are read by the Bactec® system to detect the production or absence of $^{14}C-CO_2$. Susceptibility is determined by failure of the test organism to produce $^{14}C-CO_2$ in the presence of an antimicrobial when compared to growth (presence of $^{14}C-CO_2$) in the control vial. The following formula is used to interpret results:

Growth index (GI) control > GI antimicrobial = susceptible
GI control < GI antimicrobial = resistant

Recent studies have shown the Bactec® method to be equivalent to the 1% proportion method for performing antimicrobial susceptibility for *M. tuberculosis.*[7,12] The time savings for susceptibility testing using the Bactec® is significant. The mean times required for recovery and subsequent susceptibility testing using the Bactec® and conventional method were 18.0 and 38.5 days, respectively.

It was also determined that the radiometric method gave results that were not compatible with those given using the 1% proportion method when mycobacteria other than *M. tuberculosis* were tested. Further studies need to be completed to determine if the Bactec® system can be used for antimicrobial susceptibility testing of other mycobacteria.

VII. SUMMARY

The Bactec® system certainly appears to be a rapid sensitive method for the detection of mycobacteria in clinical specimens. When studies are completed, the recommendation will likely be that 0.5 m*l* of specimen be added to each vial containing an antibiotic supplement. Tentative data show that this will make the recovery rate for one Bactec® vial equivalent to a combination of conventional media. Antimicrobial susceptibility testing for *M. tuberculosis* already appears adequate using the Bactec® system. The combination of the rapid detection time, differentiation of *M. tuberculosis* using the NAP test, and antimicrobial susceptibility testing make the Bactec® system a new revolution in mycobacteriology for the diagnosis of tuberculosis. Further development is necessary to provide rapid identification method for the MOTT group when recovered by the Bactec® system.

REFERENCES

1. Deland, F. H. and Wagner, H. N., Early detection of bacterial growth with carbon-14 labeled glucose, *Radiology*, 154, 92, 1969.
2. Cummings, D. M., Ristroph, D., Camargo, E. E., et al., Radiometric detection of the metabolic activity of *Mycobacterium tuberculosis*, *Nucl. Med.*, 16, 1189, 1975.
3. Mitchison, D. A., Allen, B. W., Carol, L., et al., A selective oleic acid albumin agar medium for tubercle bacilli, *J. Med. Microbiol.*, 5, 165, 1972.
4. Mitchison, D. A., Allen, B. W., and Lambert, R. A., Selective media in the isolation of tubercle bacili from tissues, *J. Clin. Pathol.*, 26, 250, 1973.
5. McClatchey, J. K., Waggoner, R. F., Kanes, W., Cernich, M. S., and Bolton, T. L., Isolation of mycobacteria from clinical specimens by use of selective 7H11 medium, *Am. J. Clin. Pathol.*, 65, 412, 1976.
6. Middlebrook, G., Reggiardo, Z., and Tiggert, W. D., Automatable radiometric detection of growth of *Mycobacterium tuberculosis* in selective media, *Am. Rev. Respir. Dis.*, 115, 1066, 1977.
7. Roberts, G. D., Goodman, N. L., Heifers, L., Larsh, H. W., Linder, T. H., McClatchy, J. K., McGuinis, M. R., Siddiqi, S. H., and Wright, P., Evaluation of the BACTEC radiometric method for recovery of mycobacteria and drug susceptibility testing of *Mycobacterium tuberculosis* from acid-fast smear-positive specimens, *J. Clin. Microbiol.*, 18, 689, 1983.
8. Morgan, M. A., Horstmeier, C. D., DeYoung, D. R., and Roberts, G. D., Comparison of a radiometric method (BACTEC) and conventional culture media for recovery of mycobacteria from smear-negative specimens, *J. Clin. Microbiol.*, 18, 384, 1983.
9. Morgan, M. A., DeYoung, D. R., and Roberts, G. D., Comparison of three culture methods for the radiometric detection of mycobacteria from clinical specimens, Abstr. Annu. Meet. Am. Soc. Microbiol., C222, p. 348, 1983.
10. Laszlo, A. and Eidus, L., Test for differentiation of *M. tuberculosis* and *M. bovis* from other mycobacteria, *Can. J. Microbiol.*, 24, 754, 1978.
11. Morgan, M. A., Koerr, K. A., Hempel, H. O., Goodman, N. L., and Roberts, G. D., Evaluation of the p-nitro-α-acetylamino-β-hydroxypropiophenone differential test for identification of *Mycobacterium tuberculosis* complex, *J. Clin. Microbiol.*, 21, 634, 1985.
12. Hempel, H. and Goodman, N. L., Identification of the *Mycobacterium tuberculosis* complex by the NAP system, Abstr. Annu. Meet. Soc. Microbiol., C212, p. 347, 1983.
13. Linder, T. H. and Larsh, H. W., Rapid presumptive identification of mycobacteria by the NAP-test, Abstr. Annu. Meet. Am. Soc. Microbiol., C213, p. 347, 1983.
14. Laszlo, A. and Siddiqi, S. H., Evaluation of a rapid radiometric differentiation test for the *Mycobacterium tuberculosis* complex by selective inhibition with p-nitro-α-acetylamino-β-hydroxypropiophenone, *J. Clin. Microbiol.*, 19, 694, 1984.
15. Siddiqi, S. H., Libonati, J. P., and Middlebrook, G., Evaluation of a rapid radiometric method for drug susceptibility testing of *M. tuberculosis*, *J. Clin. Microbiol.*, 12, 908, 1981.

Instrument Methods in Current Use for Identification of Microorganisms Recovered from Body Fluids

Chapter 5

OVERNIGHT AUTOMATED IDENTIFICATION SYSTEMS

Patrick R. Murray

TABLE OF CONTENTS

I. INTRODUCTION

Conventional methods for the identification of microorganisms have been automated with the development of agar replica systems and microdilution systems. Despite the fact that these systems must be incubated for the same period as conventional biochemical identification tests, they have a number of advantages including: (1) low cost per individual identification test, (2) standardization of test inoculation and interpretation, and (3) improved accuracy of identification. These different systems are analyzed in the following sections and summarized in Table 1.

II. TEST SYSTEMS

A. Agar Replica Systems

Agar replica systems are very popular in Canada, but they have not been readily adopted in the U.S. The major disadvantages with agar replica systems have been the need to prepare the biochemical supplemented agar plates, to test many organisms for the method to be cost-effective, and to incubate the tests overnight. However, for a laboratory with media preparation facilities and a large number of daily tests, this method is a very attractive alternative to conventional biochemical methods.

Repliscan® is a semiautomated system designed originally by Cathra International (Ontario, Canada) and now marketed by Diagnostic Equipment, Inc. (DEI, St. Paul, Minn.) for the identification of Gram-negative bacilli. The system consists of biochemical agar media, a multiple inoculum replicating device, a viewing-recording platform, and a computer terminal. A standardized inoculum suspension is manually transferred by the user to an individual well in the replicating device which can be used to simultaneously inoculate 36 organisms onto any number of plates containing different biochemical substrates. The Repliscan® system for gram-negative identifications uses 16 biochemical plates. After overnight incubation of the tests, the agar plates are arranged on the viewing-recording platform. The test reactions for an individual isolate are visible through viewing ports and the test reactivity is manually recorded by the user with the aid of a light pen and photosensitive detector. The computer then displays the test results in an octal code and the code interpretation. If additional tests are required for identification, a coded note is shown on the CRT screen. After the test results and identification have been recorded for an individual isolate, the viewing platform automatically moves to the next viewing position. Thus, all 36 organisms can be sequentially examined and identified.

Only limited evaluations of this system have been published. Brown and Washington[1] reported in 1978 that the Repliscan® system was able to identify 91.0% of the fermentative Gram-negative bacilli but only 73.9% of the nonfermentative organisms. *Providencia, Enterobacter, Yersinia,* and *Acinetobacter* species were frequently misidentified. Although the selection of biochemical tests in the system is not designed for the identification of oxidase-positive nonfermenters, 98% of the isolates of *Pseudomonas aeruginosa* were identified correctly. *P. maltophilia* was the only other *Pseudomonas* species that was tested, so it is not known if the Repliscan® system can differentiate among species of the *P. fluorescens* group.

Woolfrey and co-workers evaluated two versions of the Repliscan® system. In 1981 they tested 1013 isolates of Enterobacteriaceae and reported only a 62% correlation between the results of Repliscan® and the API® 20E.[2] Identification problems were encountered with *Providencia* and *Enterobacter* as was seen by Brown and Washington, as well as with *Serratia, Shigella,* and *Salmonella.* In a subsequent report,[3] they tested 692 isolates of Enterobacteriaceae in a modified system, the Repliscan® II, and correctly identified 95.2% of the isolates, including most strains of *Shigella, Salmo-*

Table 1
INSTRUMENTATION DEVELOPED FOR OVERNIGHT IDENTIFICATION SYSTEMS

	Inoculation	Test interpretation	Calculation of identification	List price* ($)
Repliscan® system				
Inoculum replicating device	Manual	—	—	32,500
Viewing-recording platform with computer terminal	—	Manual	Automated	
Gibco Sensititre® system				
Auto-Inoculator	Automated	—	—	12,000
Semiautomated viewer with DEC® Pro 350 computer-data mangement system	—	Manual	Automated	12,000
Auto-Reader with computer	—	Automated	Automated	25,000
BBL Sceptor® System				
Preparation station	Automated	—	—	7,500
Reader-recorder with DEC® PRO 350	—	Manual	Manual	3,900
computer-data management system	—	Manual	Automated	11,000
Micro-Media® systems				
Micro-Coder II	Manual	Manual	Automated	10,000
with Apple® computer	Manual	Manual	Automated	18,000
American Micro Scan®				
TouchSCAN-SR with IBM® PC XT computer	—	Manual	Automated	22,900
AutoSCAN®-4 with IBM® PC XT computer	—	Automated	Automated	39,500
PASCO® MIC/ID System				
Scanner with NCR® Worksaver 300 computer	Manual	Manual	Automated	25,000
API® UniScept System				
140 Inoculator	Manual	—	—	1,500
Autoinoculator	Automated	—	—	2,500
Touchreader with IBM® PC	—	Manual	Automated	11,000
Autoreader with				
IBM® PC	—	Automated	Automated	17,500
DEC® VT103 Computer	—	Automated	Automated	22,000
Microbiology Data System	—	Automated	Automated	50,000

* All prices were provided by the manufacturer. Discounted prices or lease arrangement are not included.

nella, and *Serratia*, and 85.2% of the *Enterobacter* isolates. The major improvements in the system were standardization of media preparation and revisions of the database in the system.

Uhl and Anhalt[10] at the Mayo Clinic developed their own version of a replica plating identification system. The system, computer-aided replica plating or CARP, utilizes an IBM® PC. Multiple organisms are inoculated onto individual biochemical plates in a manner similar to that with Repliscan®. After overnight incubation, the plates are placed on a computer screen overlay. Function keys are then used to record the biochemical test results. After all tests are entered into the computer for each organism tested, the specially designed computer program provides the identification of the tested organisms. The results of a total of 18 biochemical tests and 99 organisms can be stored in the computer before the identifications are recalled. Uhl and Anhalt reported that the CARP system correctly identified 99.7% of 588 gram-negative bacilli. This remarkable accuracy will have to be confirmed by independent investigators. The major advantages of the CARP system in comparison with the Repliscan® system are the speed of data entry and retrieval (reported to be twice as fast with the CARP system) and reduced cost of the instrumentation (approximately $4000 for the IBM PC vs. $32,500 for the DEI viewing table and computer terminal.

B. Gibco Sensititre® System

The Sensititre® system (Gibco Life Technologies, Inc.) is designed to identify 60 species of gram-negative bacilli, including the major species of Enterobacteriaceae, *Pseudomonas, Acinetobacter, Flavobacterium, Vibrio, Plesiomonas,* and *Aeromonas.* The identification panel is a 96-well microdilution tray divided into 4 24-well sections, with each well containing dried biochemical reagents. Thus, four organisms can be identified with each tray. In addition to the biochemical panels, the Sensititre® automated microbiology system includes an Auto-Inoculator, semiautomated viewer, and computer system (Digital® Professional 350 computer with a 10-megabyte hard disk, monitor, keyboard, and printer). A totally automated reader, the Auto-Reader, has also been developed for the Sensititre system which uses fluorescence technology to interpret Sensititre® susceptibility panels 5 hr after they have been inoculated. It is anticipated that the Auto-Reader system will be used with the identification panels by the end of 1985.

The test organisms are inoculated into the wells containing dried biochemical substrates with the aid of the manual pipette inoculator or the Sensititre® Auto-Inoculator. The Auto-Inoculator automatically dispenses 50 or 100 $\mu\ell$ of broth into each well of the test plates in less than 30 sec. After the panel is inoculated, sterile mineral oil is added to 5 of the 24 wells (decarboxylase base control, lysine decarboxylase, ornithine decarboxylase, urea, and H_2S), the panel is sealed with a transparent adhesive tape, and incubated overnight. Reagents must be added for five tests (nitrate, oxidase, indole, Voges-Proskauer, and tryptophan deaminase) before the panels are read with the semiautomated viewer. With the aid of the semiautomated viewer, the technologist enters test results into the computer by following the light pattern displayed in the viewer. The test reactions are then interpreted by the computer system, a seven-digit biocode is calculated, and the identification printed. A completely automated reader, the Auto-Reader, is also currently available for antimicrobial susceptibility tests. Fluorescent-labeled substrates are added to the test inoculum at the time the antibiotics are rehydrated. If the organism is resistant to an antibiotic, the labeled substrate will be utilized during growth of the organism and the Auto-Reader will detect the released fluorescent label after 5 hr of incubation. If the organism is inhibited by the antibiotic, then no fluorescence will be detectable. Gibco is currently evaluating the feasibility of this system for bacterial identifications.

The only published evaluation of the Sensititre® Gram-Negative ID panel has been performed by Staneck and co-workers.[4] They found that the system correctly identified 91.2% of the isolates of Enterobacteriaceae that they tested initially and 94.6% of all isolates with repeat testing. After this study was performed, the database was revised so the test accuracy is likely to be improved. However, further studies will be needed to confirm this.

C. Sceptor® System

The Sceptor® system (BBL Microbiology Systems, Cockeysville, Md.) consists of four identification-susceptibility testing combination panels: Gram-Negative MIC/ID, Gram-Negative Urine MIC/ID, Gram-Positive MIC/ID, and Anaerobe MIC/ID. The panels are divided into 12 rows of 7 wells with dried reagents in the individual wells. Biochemical reagents are placed in 24 wells and antibiotics in the remaining 60 wells.

In addition to the four biochemical-susceptibility testing panels, the Sceptor® system consists of an automated preparation station and a semiautomated reader-recorder, and a DEC® Pro 350 computer-data management system (similar to the system used with the Gibco Sensititre® System).

Identification tests are performed by filling the disposable reservoir in the preparation station with the test inoculum, which is then automatically dispensed into the 84 individual wells in less than 1 min. The microprocessor in the dispenser controls the volume of the inoculum (100 $\mu\ell$) and automatically turns the instrument off at the completion of the cycle. After overnight incubation, the panels are placed in the reader-recorder and the technologist manually enters the results into the instrument. The test results are then printed onto a report form and the user converts the results into a seven-digit identification profile number and the identification is determined by referring to the Sceptor® identification taxonomy guide. Alternatively, the reader-recorder can be interfaced with the Sceptor® computer-data management system which will analyze the identification results and print the organism identification.

The database for the gram-negative panels includes 72 species of oxidase-positive and oxidase-negative organisms such as members of the family Enterobacteriaceae and the genera *Acinetobacter, Aeromonas, Alcaligenes, Bordetella, Moraxella, Flavobacterium, Pasteurella, Plesiomonas, Pseudomonas,* and *Vibrio*. A total of 22 species of *Streptococcus*, 18 species of *Staphylococcus* and *Micrococcus*, and *Listeria* are included in the gram-positive panel database. The anaerobe panel can be used to identify 10 species of cocci, 17 species of gram-negative bacilli, and 28 species of gram-positive bacilli.

After inoculation of Sceptor® panels, sterile mineral oil is added to five wells in the gram-negative panels (urea, lysine decarboxylase, ornithine decarboxylase, arginine dihydrolase, and H_2S) and after overnight incubation, biochemical reagents are added to two wells in the panel (indole and tryptophan deaminase). Sterile mineral oil is added to three wells in the Gram-Positive MIC/ID panel (glucose, urea, and arginine dihydrolase) and after incubation for 48 hr, reagents are added to three wells (Voges-Proskauer, nitrate, and hippurate). Sterile mineral oil is not added to the Anaerobe MIC/ID panel, but after 48 hr of incubation in an anaerobic atmosphere, reagent is added to the indole well.

Published evaluations are only available for the gram-negative panels. Woolfrey and associates[5] reported that the Sceptor® system correctly identified 93.4% of 678 Enterobacteriaceae isolates. The majority of incorrect identifications were due to reactions with carbohydrates, and these problems were corrected when the organisms were retested. Two additional evaluations[11,12] have been presented at national meetings which reported that between 91 to 95% of the gram-negative bacilli that were tested were identified correctly. Data from a four laboratory collaborative study was also available from the manufacturer that documented 91 to 97% of isolates were identified correctly.

Evaluations of the Gram-Positive MIC/ID and Anaerobe MIC/ID panels have not been published to date. However, a preliminary multicenter study of the anaerobe panel demonstrated that 89% of 177 isolates were identified correctly (data obtained from manufacturer). These results are very impressive in comparison with other commercial anaerobe identification systems, but must be confirmed by additional studies.

D. Micro-Media® Systems

Micro-Media® (Potomac, Md.) has developed prefilled, frozen microdilution panels for identification of gram-negative and gram-positive organisms. Three identification panels are produced: a 150-well Gram-Negative UTI MIC/ID panel for quantitative antimicrobial susceptibility testing and identification of gram-negative bacilli (including urinary tract isolates), a 118-well Gram-Positive Plus MIC/ID panel for the quantitative susceptibility and presumptive identification of gram-positive cocci, and a 70-well Twin GNI panel for the simultaneous identification of two isolates of gram-negative bacilli. The only instrument used with the Micro-Media® system is the semiautomated Micro-Coder II that is designed to facilitate manual reading of the panels and automatic generation of identification reports. Inoculation of the panels is performed manually.

The biochemical tests in the 150-well Gram-Negative UTI MIC/ID and the 70-well Twin GNI panels are identical. There are 33 biochemical tests and 2 controls used to identify 68 species of fermentative and nonfermentative gram-negative bacilli. There are 20 biochemical tests incorporated in the test panel and 1 off-panel test (oxidase) used to identify members of the family Enterobacteriaceae. There are 3 supplemental tests (fermentation of dulcitol and trehalose and inhibition by colistin) included in the panel to confirm or further define an identification. Other gram-negative bacilli are identified with 18 tests incorporated in the panel and 3 off-panel tests (oxidase, growth on MacConkey agar, and motility). Supplemental tests are also included in the panel (10% lactose, nitrite production, growth in 10% NaCl broth, and inhibition by penicillin G or kanamycin). The Gram-Positive Plus MIC/ID panel can only be used for the preliminary identification of gram-positive cocci. Staphylococci, micrococci, and streptococci are differentiated using 11 biochemical tests included in the panel and 3 off-panel tests (catalase, hemolysis, and coagulase).

The test panels are inoculated with the disposable multipin inoculator which transfers the inoculum to all the wells simultaneously. Mineral oil is used to overlay eight wells in the gram-negative panels (glucose, H_2S, lysine decarboxylase, ornithine decarboxylase, arginine dihydrolase, urea, lysine control, and decarboxylase control) and two wells in the gram-positive panel (OF-glucose and arginine dihydrolase). After incubation for 24 to 48 hr, reagents are added to 4 wells in the gram-negative panels (Voges-Proskauer, tryptophan deaminase, indole, and nitrate) and 2 wells in the gram-positive panel (catalase and hippurate). The test panels are then placed onto a lighted viewing area in the Micro-Coder II and the user moves a light-emitting diode (LED) to each positive reaction and manually enters the results. Off-panel and supplemental on-panel tests are also entered into the system, and then the instrument calculates and displays the biocode number and prints the test results. Additional tests are recommended if the biochemical tests do not discriminate between multiple identifications.

Although no evaluations of the current panels have been published, the biochemical tests and database used for identification of Enterobacteriaceae are the same as were used in the previous version of this test, the MMS Quad panels. In an evaluation with 200 isolates of Enterobacteriaceae, the Quad panels were used to identify correctly 93.5% of the isolates.[6] In a second study, the Quad panels were compared with the API® 20E system. The correct identification was obtained for 87% of the isolates with the Quad system, which was similar to the results with the API® 20E system.[7] Despite the fact that the same biochemical tests and database are used in these gram-negative identification systems, the volume of broth used to inoculate the tests has been reduced from 100 μl to 40 μl and the effect of this change must be assessed. Furthermore, the accuracy of the Gram-Positive Plus MIC/ID panel must be evaluated.

E. American Micro Scan®

American Micro Scan® (Campbell, Calif.) has developed six prefilled, frozen microdilution panels for identification of gram-negative and gram-positive bacteria: five combination panels for identification and susceptibility tests (three for gram-negative enteric bacilli including urine isolates, one from gram-positive cocci, and one for anaerobes) and one panel for identification of gram-negative bacilli only. These panels are inoculated with a disposable multipin inoculator which transfers 5 $\mu\ell$ to each well, and the interpretation and recording of the test results are performed with the aid of either a semiautomated (TouchSCAN-SR) or totally automated (AutoSCAN-4) instrument. American Micro Scan® has also developed a dried panel (HN-ID panel) for the identification of *Haemophilus, Neisseria, Branhamella,* and *Gardnerella.* This HN-ID panel can also be interpreted with the TouchSCAN-SR or AutoSCAN-4 instruments, American Micro Scan® anticipates developing other dried panels which would circumvent the storage problems associated with the frozen panels.

The gram-negative panels are used to identify 72 fermentative and 30 nonfermentative organisms. The identification profiles are determined with 23 tests, 10 of which are unique to each group of organisms and 13 tests are common to both groups. A total of 43 species of gram-positive bacteria, including species of *Streptococcus, Staphylococcus, Micrococcus,* and *Listeria,* are identified with the Gram-Positive Combo panel, 64 species of anaerobes with the Anaerobe Combo panel, and 20 species of bacteria with the HN-ID panel.

After the inoculum is manually transferred into the gram-negative panel, sterile mineral oil is added to seven wells (glucose, urea, H_2S, lysine decarboxylase, ornithine decarboxylase, arginine dihydrolase, and decarboxylase control) and nine wells are covered with a sealing strip. Two wells in the Gram-Positive Combo panel are overlayed with mineral oil (urea and arginine dihydrolase). No well in the Anaerobe Combo is overlayed with oil and only one well (urea) in the HN-ID panel is overlayed. Panels for gram-positive bacteria and fermentative gram-negative bacilli are incubated for 24 hr and tests for the identification of most nonfermentative gram-negative bacilli and all anaerobes are incubated for 48 hr. The HN-ID panel is incubated for only 4 hr. Before the panels can be interpreted, reagents are added to four wells (Voges-Proskauer, tryptophan deaminase, indole, and nitrate) in the gram-negative panels, to four wells (Voges-Proskauer, nitrate, hippurate, and β-lactamase) in the gram-positive panel, and to six wells (nitrate, nitrite, indole, and three peptidase reactions) in the HN-ID panel. In addition to the reagents added to the esculin, nitrate, and indole wells in the Anaerobe Combo panel, brom-cresol purple is added to the 15 carbohydrate wells. Thus, reagents and indicator are added to 18 identification test wells in the anaerobe panel.

If the TouchSCAN-SR is used for interpretation of the tests, the technologist inserts the panel into a holder which illuminates the biochemical wells. The technologist then interprets each reaction visually and enters the positive or negative results by the aid of a probe and sonic reader. The probe activates an audible signal which is identified by two microphones in the sonic reader. This signal is then relayed to the system's IBM® computer which compiles the test reactions and generates a code number and identification. The use of the sonic reader is a novel approach to panel reading systems. The advantage of this system in comparison with the previous TouchSCAN instrument is that an overlay template is not needed and any panel size can be used. The manufacturer reports that background electrical or audible noise does not interfere with test interpretation, but this will have to be corroborated by independent investigators.

If the American Micro Scan® AutoSCAN-4 is used, the test panel is inserted into the instrument and then transmission of a tungsten-halogen white light is measured simultaneously in each of the 96 wells. These readings are repeated at 6 different wavelengths and are completed in 6 sec. The electrical signals are then converted into a digital reading and the results of the individual test reactions are interpreted at predetermined wavelengths. The analysis at multiple wavelengths permits better resolution of borderline reactions. The results are then automatically transferred to the IBM® computer which prints the identification results.

A prototype modification of the AutoSCAN (the AutoSCAN-W/A or Walk Away system) has also been developed. This system is purported to offer total automation. After the panels are manually inoculated, they are inserted into the instrument and incubated overnight. Thereafter, the instrument adds reagents and interprets the panels automatically. The date when this totally automated system will be available has not been determined.

Extensive evaluations of the American Micro Scan® Gram-Negative Combo and Urine Combo panels have demonstrated that the panels can be used to identify between 93 and 98% of isolates of Enterobacteriaceae.[5,8] Identification of non-Enterobacteriaceae has not been as good with only 82% of the isolates identified correctly in one study.[13] However, all isolates of *Pseudomonas aeruginosa, Acinetobacter, Moraxella,* and *Alcaligenes* were identified accurately. The accuracy of the Gram-Positive Combo panel has not been determined. Savuto and Ellner,[9] in their evaluation of the Anaerobe Combo panel, reported that 84% of their tested organisms were able to grow in the system and 78% of these isolates were identified correctly at the species level.

F. PASCO® MIC-ID System

PASCO® Laboratories, Inc. (Denver, Colo.) has developed a frozen microdilution susceptibility-identification system for gram-negative bacilli. Two 104 well panels are available; a Gram-Negative MIC/ID Combo panel and a Cystitis MIC/ID Combo panel. The panels have 30 wells for biochemical tests, with 27 tests used for identification of oxidase-negative bacilli and 27 tests used for oxidase-positive bacilli. The manufacturer claims that 134 species of bacilli can be identified with this system although this will have to be corroborated in clinical trials. Reading and interpretation of the PASCO® MIC/ID system is performed with the aid of the semiautomatic PASCO® scanner.

Inoculum is introduced manually into the panels by use of a disposable, multipin inoculator. Mineral oil is then added to 5 wells (lysine decarboxylase, arginine decarboxylase, ornithine decarboxylase, urea and OF-glucose) and the panels are incubated for 18 to 24 hr. Before the panels can be read, reagents are added to four wells (nitrate, indole, trytophane deaminase, and Voges-Proskauer). The test reactions are interpreted by inserting the panels into the PASCO® scanner and then, with the aid of a light pen and LED indicator lights, the technologist manually enters the results. After results for all test panels are entered into the system, the NCR® Worksaver 300 computer analyzes the test reactions and prints the organism identification.

The identification capabilities of this system have been recently developed, so clinical studies of the accuracy of the system have not been reported. These studies will have to be performed before any assessment of the utility of the system can be made. In addition to the susceptibility and identification capabilities of this system, the NCR® computer with its 80-286 microprocessor and 10-megabyte hard disk has stand-alone multiuser capabilities. Thus, it can be used for multiple laboratory applications.

G. API® UniScept System

The UniScept System (Analytab Products, Inc., Plainview, N.Y.) is an automated approach for using the traditional API® identification products. Specifically, the following special identification strips with transparent backing can be used with the UniScept system: API® 20E (for identification of enteric gram-negative bacilli), API® 20GP (for gram-positive cocci), and An-Ident (for anaerobes). Whereas the range of species that can be identified with the API® 20E and An-Ident is extensive, the number of species identified with the API® 20GP is restricted to 14 species of staphylococci and four species of group D streptococci.

The UniScept system consists of three components: an inoculator, a reading device, and a computer. Two inoculators, the semiautomated 140 Inoculator and the fully automated Autoinoculator, and two readers, the Touchreader and the automated Autoreader, are available with the UniScept system. The 140 Inoculator transfers inoculum from a seed tray to the biochemical wells by means of a disposable multitip pipette. The microprocessor-controlled Autoinoculator is programmed to deliver specific volumes of inoculum to the different wells in the test panels. For example, the system delivers 50 μl to microdilution susceptibility wells, 100 μl to the KB susceptibility wells and most biochemical wells, and 300 μl to specified biochemical wells. The inoculum is automatically drawn into a syringe and then dispensed into the test wells. It is anticipated that this inoculator will soon be used to dispense sterile mineral oil into the appropriate wells. One of the readers is then used to interpret the test results. The test trays can be placed on the Touchreader and results manually read and entered into the instrument by the technologist using a keyboard panel and LED indicator lights. The Touchreader can also be interfaced with either a printer or the UniScept microbiology data system. The second reader, the Autoreader, is an automated, multi-wavelength, multiposition instrument. After the test tray is placed in the instrument, each well is examined at as many as five different wavelengths and at multiple positions in the well. This approach facilitates discrimination of weak reactions and reactivity at only one part of the test well, e.g., top of indole reaction well, bottom of carbohydrate well. The Autoreader can be interfaced with the user's personal computer or with the UniScept microbiology data system, which is a DEC® VT103 computer with a 20-megabyte hard disk and printer.

Extensive evaluations of the API® enteric identification systems have been published and have demonstrated a high degree of accuracy. However, reports have not been published that examine these tests in the semiautomated and fully automated UniScept systems. Likewise, the API® 20GP and An-Ident must also be evaluated more extensively in this system.

H. Aladin System

The Aladin system (Analytab Products, Inc., Plainview, N.Y.) is a further refinement of the UniScept concept. At the present time only a prototype instrument has been developed which features totally automated operation after the inoculum is prepared for identification and susceptibility tests and complete flexibility in the selection of various combinations of tests. All of the API® identification tests that are used in the UniScept system can be used in the Aladin including 20E, 20GP, and An-Ident, as well as other API® tests such as the Rapid E, Rapid NFT, and Yeast IDent tests. The ability to use 4- to 5-hr tests such as the Rapid E and An-Ident expands the utility of this system beyond overnight testing.

The Aladin system consists of four components contained within one structure: an incubator, reader station, reagent-dispensing system, and waste-disposal area. The incubator contains 60 test slots divided into 2 rows of 30. Into each slot is placed a universal carrier which can hold one identification system and, when desired, an antimicrobial susceptibility test. After the universal carrier is loaded into the system, the elevator and mechanical arm moves the carrier into the reader station where the specimen demographic information and test product (e.g., type of identification and susceptibility test) identification is recorded. An initial reading of the individual wells is determined (T_0 reading) and then the carrier is moved back into the incubation chamber. After the prescribed incubation period for the individual tests, the carrier is moved back into the reader for a final reading. The T_0 and T_{final} readings are recorded by the video image processing system in the reader which consists of one of two cameras with a fixed focal length and a color wheel positioned above the camera. One camera is available for each of 30 test slots. The camera is programmed to focus on the area of interest for a specific well (e.g., top of indole well or bottom of fermentation wells) where color changes are anticipated. The color wheel is then rotated and a reading is recorded at each wavelength. The software within the system then selects the reading (or combination of readings) that are relevant for each well and records the results as positive or negative. The system can also discriminate between areas of weak and strong activity. If reagents are required before the tests are initially incubated or before the final reading, then the carrier is transported to the reagent dispensing system where the appropriate reagents are added. If the tests must be incubated after the reagents are added and before the final readings are performed, then this is automatically performed. The Aladin is a continuous monitoring system which discards all tests as they are completed and permits the insertion of a new carrier into the vacated slot. In addition, off-line incubation is possible with the system, which is particularly useful for tests that require prolonged incubation before the final reading is performed.

The anticipated release date for this system is in 1986. It is expected that by that time clinical evaluations of the system will be available. At the present time, the Aladin system should be considered to be a very flexible, sophisticated system that awaits clinical studies to assess its role in the clinical laboratory.

III. CONCLUSIONS

Microdilution identification systems were initially introduced in the mid-1970s and have proven to be very useful for laboratories that combine the tests with quantitative microdilution antimicrobial susceptibility testing. The biochemical tests are available in either the frozen state (which are generally simple to inoculate manually but impose a transport and storage requirement) or in the dried state (which require precise inoculation but are easy to store). Instrumentation has been developed that facilitates the inoculation of these tests and interpretation of the test results. However, each of these improvements are introduced at a significant cost and the tests must still be incubated as long as conventional biochemical tests. Furthermore, only limited evaluations of these tests have been published to date, which is surprising when one considers the length of time these tests have been used. It is important that these tests be more critically evaluated before they are accepted by the microbiology community. In addition, more effort should be directed towards developing rapid, same-day test results. The feasibility of this approach has been demonstrated by Sensititre® (e.g. Auto-Reader) and American Micro Scan® (e.g., HN-ID panel).

REFERENCES

1. Brown, S. D. and Washington, J. A., II, Evaluation of the Repliscan system for identification of Enterobacteriaceae, *J. Clin. Microbiol.*, 8, 695, 1978.
2. Woolfrey, B. F., Quall, C. O., Fox, J. M., Evaluation of the Repliscan System for Enterobacteriaceae identification, *J. Clin. Microbiol.*, 13, 58, 1981.
3. Woolfrey, B. F., Fox, J. M. K., Quall, C. O., Evaluation of the Repliscan II System for identification of Enterobacteriaceae, *J. Clin. Microbiol.*, 14, 408, 1981.
4. Staneck, J. L., Vincelette, J., Lamothe, F., and Polk, E. A., Evaluation of the Sensititre System for identification of Enterobacteriaceae, *J. Clin. Microbiol.*, 17, 647, 1983.

5. Woolfrey, B. F., Lally, R. T., and Qually, C. O., Evaluation of the Auto Scan-3 and Sceptor Systems for Enterobacteriaceae identification, *J. Clin. Microbiol.*, 17, 807, 1983.
6. Borchardt, K. A. and Gibson, J., Comparison of enteric identification systems, *Health Lab. Sci.*, 14, 5, 1977.
7. Barry, A. L., Badal, R. E., and Effinger, L. J., Identification of Enterobacteriaceae in frozen microdilution trays prepared by Micro-Media Systems, *J. Clin. Microbiol.*, 10, 492, 1979.
8. Mathewson, J. J., Simpson, R. B., and Brooks, F. L., Evaluation of the Micro Scan Urinary Combo Panel and API 20E System for identification of glucose-nonfermenting gram-negative bacilli isolated from clinical veterinary materials, *J. Clin. Microbiol.*, 17, 139, 1983.
9. Savuto, P. S. and Ellner, P. D., Evalution of a microtiter system for identification of anaerobic bacteria, *J. Clin. Microbiol.*, 20, 81, 1984.
10. Uhl, J. R., and Anhalt, J. P., Identification and susceptibility testing using an IBM PC, Abstr. Annu. Meet. Am. Soc. Microbiol., American Society for Microbiology, Washington, D.C., 1984, Abstr. C221.
11. Cohen, R. L., Identification of aerobic Gram-negative bacilli using the BBL Sceptor system, Abstr. Annu. Meet. Am. Soc. Microbiol., American Society for Microbiology, Washington, D.C., 1982, Abstr. C152.
12. Johnson, K. S., Raymon, J. L., and Gerlach, E. H., Simultaneous biochemical identification and antimicrobial susceptibility testing: evaluation of the Sceptor system, Abstr. Annu. Meet. Am. Soc. Microbiol., American Society for Microbiology, Washington, D.C., 1983, Abstr. C345.
13. Rhoden, D. L., Esaias, E. L., Baker, C. N., Helsel, L. O., Smith, P. B., and Thornsberry, C., Evaluation of the MicroScan system for the identification of Gram-negative bacilli, Abstr. Annu. Meet. Am. Soc. Microbiol., American Society for Microbiology, Washington, D.C., 1983, Abstr. C183.

Chapter 6

RAPID AUTOMATED IDENTIFICATION SYSTEMS

Patrick R. Murray

TABLE OF CONTENTS

I. INTRODUCTION

The careful selection of biochemical tests, development of extensive databases, and the objective measurement of test reactions with automated readers have ushered in a new generation of instruments capable of identifying microorganisms after a few hours of incubation. The advantages of these systems are obvious. The rapid identifications of organisms is compatible with rapid antimicrobial susceptibility testing. Even in laboratories that do not perform rapid susceptibility tests, the availability of prompt identification results can be used for the selection of better defined empiric antibiotic therapy. These identification systems can also improve test turnaround time, as well as automate and standardize the interpretation of the test results. The computer-assisted analysis of test results should eliminate technical error and test bias. Finally, automation of costly conventional biochemical tests should reduce the cost associated with identification of microorganisms. The following is an analysis of systems that can be used to identify bacteria and yeasts significantly more rapidly than with conventional methods.

II. AUTOMATED SYSTEMS

A. Vitek® System

The Vitek® system (formerly called the AutoMicrobic® or AMS® system; Vitek Systems, McDonnell-Douglas Health Systems Co.) was one of the first truly automated identifications systems in microbiology. The original system was developed for the enumeration and identification of organisms in urine specimens. This function has been reviewed extensively[22,34] and will not be discussed here. Subsequent modifications of the system have expanded its capabilities to include susceptibility testing of gram-positive and gram-negative organisms, as well as identification of gram-positive and gram-negative bacteria and yeast.

The Vitek® system consists of a manual diluent dispenser, a filling unit which fills and then seals ten test cards simultaneously, a reader/incubator module with a capacity for 30 to 120 test cards, a computer control module with a Motorla® 68010 16/32 bit microprocessor and 10-megabyte Winchester hard disk, data terminal, and multicopy printer. A data management system can also be added to the system for epidemiologic analysis of demographic and test data. In addition to the biochemical cards that are used for enumerating and identifying urinary tract isolates, 30-well transparent test cards have been developed for the identification of gram-negative bacilli (the GNI card; formerly the EBC+ card), gram-positive cocci and bacilli (the GPI card), and yeast (the YBC card).

The gram-negative and gram-positive identification test kits were modified extensively in the spring of 1984. Selected biochemical tests were removed from the identification cards, other tests were added, and the databases were expanded. There are 29 biochemical reactions currently tested in the GNI and GPI cards and 26 reactions in the YBC card. The GNI card database includes 38 species of Enterobacteriaceae, 15 species of *Pseudomonas,* four species of *Flavobacterium,* two species each of *Achromobacter, Bordetella, Acinetobacter,* and *Vibrio,* and one species of *Agrobacterium, Eikenella, Aeromonas, Plesiomonas,* and *Pasteurella.* The GPI card can be used to identify 23 species of *Streptococcus,* 15 species of *Staphylococcus,* four species or groups of *Corynebacteria,* and the genera *Aerococcus, Listeria,* and *Erysipelothrix. Corynebacterium* group JK cannot be identified with the GPI card. The YBC identification database includes 12 species of *Candida,* 6 species of *Cryptococcus,* 3 species of *Rhodotorula,* 4 species of *Trichosporon,* and 1 species each of *Saccharomyces* and *Geotrichum.*

Table 1
MINIMUM TIME REQUIRED FOR IDENTIFICATION OF ORGANISMS WITH THE VITEK® SYSTEM

Test card	Organism	Earliest time for identification (hr)
GNI	*Shigella sonnei*	6
	Shigella spp.	6
	S. dysenteriae	9
	Yersinia pestis	9
	Other glucose fermenters	4
	Pseudomonas aeruginosa	4
	Actinetobacter anitratus	6
	P. cepacia	6
	P. maltophilia	6
	Flavobacterium meningosepticum	9
	Pasteurella multocida	9
	Other glucose nonfermenters	18
GPI	Gram-positive bacteria	4
YBC	Yeast	24

The inoculum for the Vitek® system is prepared from either a nonselective or selective agar medium. The test organism is inoculated into sterile saline and the inoculum size is adjusted manually to the appropriate McFarland standard (no. 0.5 for gram-positive bacteria, no. 1 for Enterobacteriaceae, and no. 2 for other gram-negative bacilli and yeasts). The inoculum is pipetted into the diluent dispenser tube from which the inoculum is transferred into the appropriate identification card by the automated filling unit. The bacterial identification cards are then placed into the reader/incubator module where the cards are incubated for as long as 18 hr. At hourly intervals light transmission through the individual test wells is measured and the information is stored in the Vitek® computer module. The percent change in light transmission for each well is compared with a threshold value and when the value is exceeded, the reaction is considered positive. The YBC identification card is incubated off-line in a 30°C incubator for 24 hr. At the end of the incubation period, the card is removed from the incubator and placed in the reader/incubator module for a reading and analysis. Once the bacterial and yeast identification cards have been incubated for a sufficient period of time (Table 1), the biochemical test results are matched to the database. The first and second identification choices of the organism, as well as the normalized percent probability for the identifications, are then printed.

The Vitek® system gram-negative identification tests have undergone extensive modifications since they were first introduced. Many laboratories have evaluated the first gram-negative identification card, the EBC card, which was developed for the 8-hr identification of members of the family Enterobacteriaceae (Table 2). Between 92 and 99% of the tested isolates were identified correctly with this card, although problems were encountered with some organisms including *Klebsiella oxytoca* (not in the database), *Enterobacter agglomerans, E. cloacae, Citrobacter freundii, Morganella morganii,* and *Providencia* species. Many of these incorrect identifications were due to the limited database and aberrant reactions with H_2S, adonitol, malonate, and arginine dihydrolase. In addition, as many as 7% of the cards were reported to be filled incorrectly with the instrument.[21]

The second generation Vitek® identification card, the EBC+ card, was developed for the identification of non-Enterobacteriaceae within 18 hr, as well as the 8-hr iden-

Table 2
IDENTIFICATION OF ENTEROBACTERIACEAE WITH VITEK® GRAM-NEGATIVE ID CARDS

Vitek® identification card		Number (%) of organisms			
	Tested	Correctly identified	Equivocal identification	Incorrectly identified	Ref.
EBC	6470	6253 (96.7)	—[a]	217 (3.3)	23
EBC	192	186 (96.9)	—[a]	6 (3.1)	27
EBC	776	745 (96.0)	—[a]	31 (4.0)	16
EBC	1528	1516 (99.2)	—[a]	12 (0.8)	0
EBC	933	875 (93.8)	45 (4.8)[b]	13 (1.4)	21
EBC	1401	1320 (94.2)	57 (4.1)[c]	24 (1.7)	17
EBC	502	485 (96.6)	—[a]	17 (3.4)	13
EBC	650	599 (92.2)	28 (4.3)[b]	23 (3.5)	3
EBC+	650	585 (90.0)	45 (6.9)[b]	20 (3.1)	
EBC+	1304	1199 (92.0)	65 (5.0)[b]	40 (3.0)	5
EBC+	201	191 (92.0)	—[d]	10 (8.0)	18
EBC+	740	675 (91.2)	40 (5.4)[b]	25 (3.4)	4
EBC+	1471	1334 (90.7)	84 (5.7)[b]	53 (3.6)	6
EBC+	400	380 (95.0)	—[a]	20 (5.0)	25
EBC+	640	614 (95.9)	—[a]	26 (3.1)	42

[a] Identifications with probability <0.50 were considered equivocal; number of equivocal identifications was not stated.
[b] Identifications with probability <0.80 were considered equivocal.
[c] Identifications with probability <0.90 were considered equivocal.
[d] Identifications with probability <0.95 were considered equivocal; number of equivocal identifications was not stated.

tification of Enterobacteriaceae. Modifications of the EBC that were incorporated in the EBC+ card included the addition of glucose oxidation, acetamide, and cetrimide tests, as well as the off-line oxidase test. The accuracy of identifying Enterobacteriaceae with the EBC and EBC+ cards was essentially identical (Table 2). However, generally less than 85% of non-Enterobacteriaceae species were identified accurately with the EBC+ card with the exception of *Aeromonas hydrophila, Pseudomonas aeruginosa,* and *Acinetobacter calcoaceticus* subsp. *lwoffi* (Table 3). With both the EBC and EBC+ cards, the identification specificity was low with uncommon organisms such as *Citrobacter amalonaticus, Enterobacter agglomerans, Klebsiella* spp., *Serratia liquefaciens, Yersinia, Vibrio* spp., *Acinetobacter,* and *Pseudomonas* species other than *P. aeruginosa.* Thus, if an isolate was identified as one of these organisms, the identification was frequently incorrect. This was partially corrected by retesting equivocal reactions (e.g., identifications with a probability <0.80), although this also lowered the percentage of organisms that could be identified initially with the Vitek® system.

The EBC+ card has now been replaced by the Gram-Negative Identification (GNI) card. The cetrimide and melibiose reactions were removed from the EBC+ card, eight other tests were either modified or added, and the database was extensively updated. Although the GNI card has not been systematically evaluated at this writing, Woolfrey and associates[42] reported that many incorrect identifications observed with the EBC+ card were resolved with the GNI card. It was unclear from their study whether this improved accuracy was due to the modifications of the identification systems or simply due to retesting the isolates.

One limitation with the EBC, EBC+, and GNI identification cards is the need to incubate the cards for 8-hr for the identification of most Enterobacteriaceae (Table 1),

Table 3
IDENTIFICATION OF NON-ENTEROBACTERIACEAE WITH VITEK® GRAM-NEGATIVE ID CARDS

Vitek® identification card	Number (%) of organisms				Ref.
	Tested	Correctly identified	Equivocal identification	Incorrectly identified	
EBC+	121	117 (96.7)	—[a]	4 (2.3)	26
EBC+	472	406 (86.0)	24 (5.1)[a]	42 (8.9)	39
EBC+	271	251 (92.6)	9 (3.3)[b]	11 (4.1)	6
EBC+	252	232 (92.4)	4 (1.6)[b]	15 (6.0)	5
EBC+	102	97 (95.1)	—[a]	4 (3.9)	42

[a] Identifications with probability <0.50 were considered equivocal; number of equivocal identifications was not stated.
[b] Identifications with probability <0.80 were considered equivocal.

whereas other automated and manual same-day identification systems can generally have the results available within 4 to 5 hr. Ferraro et al.[16] reported that the Vitek® system could be used to report a presumptive identification after 4 hr of incubation for the 11 bacterial species most commonly isolated at their institution. They observed that 83% of these isolates could be correctly identified at the genus or species level. Barry and Badal[4] have also examined the accuracy of early identifications of Enterobacteriaceae and non-Enterobacteriaceae with the updated EBC+ card. They reported that the early identification of *Morganella morganii, Acinetobacter* sp., *Yersinia* spp., *Salmonella, Shigella, Enterobacter agglomerans, Pseudomonas* spp. (other than *P. aeruginosa* or *P. maltophilia*), *Klebsiella* spp. (other than *K. pneumoniae* or *K. oxytoca*), *Citrobacter amalonauticus, Serratia liquefaciens,* and *Vibrio* should be considered nonspecific responses even when the identification was associated with a high probability of accuracy. However, 96% of the other identifications obtained at 4 hr were accurate. Thus, they found that approximately 60% of the Enterobacteriaceae isolates could be identified at 4 hr if all nonspecific results and equivocal results (e.g., $p < 0.80$) were excluded.

The Vitek® GPI card can be used to identify species of *Staphylococcus, Streptococcus, Aerococcus, Listeria, Erysipelothrix,* and *Corynebacterium*. Studies have not been published that assess the test accuracy of this card with the latter two genera and only single studies have analyzed the identification accuracy with *Listeria* and *Aerococcus*. All 30 isolates of *L. monocytogenes* that were tested by Ruoff and associates[35] were identified at a high level of probability with the GPI card and 24 of 26 isolates of *Aerococcus* tested by Facklam et al.[15] were classified correctly.

The three published studies[1,19,35] that examined the test accuracy of the GPI card with *Staphylococcus* were performed before the biochemical tests in the card were changed and the database was expanded. Thus, the capabilities of the present identification system are unknown. It is clear from the early studies with the GPI card that most commonly isolated species (*S. aureus, S. epidermidis,* and *S. saprophyticus*) could be identified with an accuracy of 95% or better. However, the identification accuracy was significantly less with other coagulase-negative *Staphylococcus* species, with only 28 to 69% identified correctly. Furthermore, between 47 and 65% of the misidentified species were classified incorrectly as other *Staphylococcus* species. This poor performance was in part due to the limited database present in the first version of the Vitek® GPI system, which has now been expanded from 6 to 12 species of

Table 4
IDENTIFICATION ACCURACY WITH VITEK®
YBC CARD

No. organisms tested	No. (%) correctly identified	No. (%) with probability value ≥0.9[a]	Ref.
253	242 (95.7)	198 (78.3)	33
352	299 (84.9)	246 (69.9)	20
1095	1033 (94.3)		30

[a] Number of identifications with $p \geq 0.8$ was not stated.
[b] Number of identifications with $p \geq 0.9$ was not stated.

coagulase-negative staphylococci. In addition, novobiocin, xylose, and a modified urease test have been included in the new GPI card. These changes should correct many of the misidentifications of *S. hominis* and *S. warneri* and *S. saprophyticus* or *Staphylococcus* spp. Almeida and associates[1] reported that the average incubation time required for identification of coagulase-negative staphylococci ranged from 9.6 hr for *S. saprophyticus* to 13.0 hr for *S. epidermidis,* with a mean value of 10.8 hr.

The Vitek® GPI system can be used for an accurate identification of *Streptococcus pneumoniae* and groups A, B, and D streptococci (including both enterococcal and nonenterococcal strains). However, other strains of β-hemolytic streptococci and the viridans group pose a problem for this identification system. Appelbaum and associates[2] reported that only 58.3% of groups C, F, and G could be identified with the GPI card, and three studies documented that only 57 to 79% of the viridans group species were identified correctly.[2,15,35] Furthermore, many of these accurate identifications were at a low confidence level with a probability between 50 and 80%. The majority of the misidentifications with the GPI card were incorrect species designations rather than isolates classified as streptococcal species or as isolates that the system was unable to identify. Thus, these errors may not be apparent to the user. The average time required for an identification was 8 hr, with the rapidly growing enterococci identified as early as 4 hr but the other more slowly growing streptococcal isolates requiring as long as 15 hr for an identification.[2] This is, however, a shorter period than is required with most conventional tests.

Three laboratories have published evaluations of the Vitek® YBC card.[20,30,33] From 85 to 96% of the yeasts were identified accurately with this system, although many isolates were classified with a low identification probability (Table 4). Virtually all common clinical isolates (e.g., *Candida albicans, C. glabrata, C. tropicalis, C. krusei, C. parapsilosis,* and *Cryptococcus neoformans*) were identified correctly, whereas identification problems were encountered with the less commonly isolated species of *Candida, Cryptococcus,* and *Trichosporon.*[20,33] Oblack and associates[33] reported that whereas 96% of their isolates were correctly identified when morphology was used, this identification accuracy decreased to 84% when the biochemical tests were used alone. Thus, microscopic morphology of the isolates must be used in conjunction with the biochemical test results for an accurate identification. The incubation period for the YBC card cannot be extended to 48 or 72 hr for the identification of slowly reactive yeasts because the prolonged incubation causes additional misidentifications.[30,33]

B. MS®-2, Avantage, and Quantum II

The MS®-2 system for bacterial identifications was first marketed by Abbott Laboratories (Diagnostic Division, Dallas, Tex.) in 1979. In 1983 the Avantage system was introduced as the successor to the MS®-2 and in 1984 the Quantum II system was added as a complementary identification system. In contrast with the MS®-2 and Avantage systems, the Quantum II currently lacks the ability to be used for performing antimicrobial susceptibility tests or screening urine specimens for the presence of high grade bacteriuria.

The basic components of the MS®-2 System are the analysis module, the control module, and the identification cartridge. The analysis module is an incubator-shaker with an electro-optical scanning device. Microbial growth and utilization of biochemical substrates are monitored by photodetectors and matched light emitting diodes at a wavelength of 665 nm. Each analysis module contains eight compartments for cartridges, with the eighth compartment used for the identification cartridges. The control module contains a microprocessor that controls the operation of the analysis modules (four analysis modules can be monitored by one control module), stores data generated from the analysis modules, identifies the isolates by comparing the biochemical test results with the database in the MS®-2 system, and prints the identification result and the probability of accuracy. The identification cartridge is a disposable plastic container subdivided into 20 chambers. In the initial MS®-2 System, lyophilized biochemicals were present in 17 of the 20 chambers in the gram-negative (Enterobacteriaceae) identification cartridge and in 20 chambers in the yeast identification cartridge.

The major modification of the MS®-2 system with the development of the Avantage system was with the control module. The computer hardware and software were upgraded which expanded the identification capabilities of the system. The bacterial identification cartridge was also expanded to include 20 biochemical tests for the identification of members of the family Enterobacteriaceae as well as other common gram-negative bacilli, including *Pseudomonas aeruginosa, P. cepacia, P. maltophilia, P. fluorescens/putida* group, *Acinetobacter calcoaceticus* (var. *anitratus* and *lwoffi), Flavobacterium meningosepticum/IIb, Aeromonas hydrophila, Plesiomonas shigelloides,* and *Pasteurella multocida.* The analysis module in the Avantage system is the same as that used in the MS®-2 system.

The Quantum II is a dual-wavelength spectrophotometer with the database for identification of gram-negative bacilli and yeasts contained in an attached memory module. The current database in the Quantum II is an update version of that contained in the Avantage system.

Identification tests with the three systems are performed in basically the same manner. Bacterial suspensions are prepared in 5 mℓ of sterile distilled water and adjusted to a turbidity equivalent to a McFarland 0.5 standard. The top covering of the identification cartridges are perforated and 200 $\mu\ell$ of the standardized inoculum is dispensed into each chamber. The top of the cartridge is sealed with an adhesive covering and the cartridge is then inserted into the analysis module for the initial reading. The initial reading is not performed with the Quantum II instrument. The cartridges are then incubated at 35°C in an off-line air incubator (the incubator in the analysis module is not used for identification tests) for 4 hr for oxidase-negative organisms or 5 hr for oxidase-positive organisms. After the prescribed incubation period, indole reagent is added to the chamber containing L-tryptophan. The cartridges are then inserted into the analysis module or Quantum II system for interpretation of the individual reactions. Yeast identification tests are performed in the same manner with two modifications. The test inoculum is prepared in 0.5% Noble agar base medium and the inoculated cartridges are incubated at 30°C for 24 hr.

The first published comprehensive evaluation of the MS®-2 bacterial identification (BID) system was a three-laboratory collaborative study.[31] Of the 1154 isolates of Enterobacteriaceae that were tested, 93.5% were correctly identified, with no identification and an incorrect identification reported for 1.0 and 5.5% of the isolates, respectively. A total of 8% of the correct identifications were reported with a low likelihood of accuracy ($p < 0.8$); therefore, only 86.1% of the isolates were identified correctly with no supplementary tests required. Organisms that were frequently not identified with the MS®-2 BID system were *Enterobacter agglomerans, E. cloacae,* and *Klebsiella ozaenae.* Nonspecific identifications were observed with *Citrobacter freundii, Salmonella, K. ozaenae, Hafnia, E. agglomerans, Serratia* species, and *Providencia rettgeri.* DiPersio and associates[14] compared the original MS®-2 software with an updated version and found that the identification accuracy for 968 strains of Enterobacteriaceae was improved from 85.3 to 94.4%. The percentages of equivocal results ($p < 0.8$) and incorrect results were reduced, although specificity problems were still encountered with organisms identified as *Salmonella, K. ozaenae, E. agglomerans, Serratia liquefaciens,* and *Providencia* species. The biochemical reactions that were frequently responsible for these erroneous reports were arabinose, adonitol, mannitol, and ornithine decarboxylase.

One multicenter study has analyzed the identification accuracy of the Avantage for common oxidase-positive, nonenteric bacilli.[24] In the first phase of the evaluation with 200 challenge organisms, 95% of the isolates were classified correctly, as were 98% of the organisms not included in the database. In the second phase of the study, 95% of clinical isolates were identified correctly with identification problems encountered only with *Pseudomonas cepacia, P. fluorescens, P. putida,* and *Flavobacterium meningosepticum.*

Two laboratories have evaluated the identification accuracy of the Quantum II system. Murray et al.[32] reported that 97.0% of the Enterobacteriaceae and 83.3% of the non-Enterobacteriaceae were identified correctly. Six of the eight nonenteric isolates that were not identified with the Quantum II system were identified upon retesting. Sylvester and Washington[41] reported similar results. The slightly improved performance of the Quantum II in comparison with the Avantage is most likely due to the updated database used in the Quantum II.

Three studies have evaluated the accuracy of the Abbott Laboratories yeast identification tests. Cooper and associates[11] reported that 93% of stock cultures and 96% of clinical isolates could be identified with the MS®-2 yeast identification system if the results of the germ tube test were included. Identification problems were encountered only with the less commonly isolated organisms. They reported that microscopic morphology on corn meal agar was not required for the identification of most isolates, in contrast with the Vitek® system. Two similar studies have been performed with the Quantum II yeast identification tests.[29,36] Salkin and associates[36] reported that the overall accuracy with 239 yeast isolates was 85.8% with problems encountered again with the identification of uncommon isolates. The investigators attributed most of the identification errors to overinoculation of the cartridges (a problem that they were unable to correct even with extensive testing experience), and mechanical problems with the Quantum II instrument. They concluded that the limited database and the physical problems with inoculation and reading the test cartridges made the Quantum II system unsuitable for the routine identification of yeasts. Kiehn et al.[29] also reported that the Quantum II system was unsatisfactory for yeast identification. They were able to identify only 80.4% of 245 yeast isolates that they tested, with 7.8% of the isolates identified incorrectly and 11.8% with no identification reported. False-positive reactions were responsible for many of the erroneous identifications and were also observed with correctly identified isolates. Although not specifically determined, many of the false-positive reactions were likely to be due to problems with preparation of the test inoculum.

C. Autobac® Series II System

An innovative application of the Autobac® Series II system (General Diagnostics, Warner-Lambert Co., Morris Plains, N.J.) is used for bacterial identification. Whereas other identification systems measure the metabolic activity of organisms with various substrates, the Autobac® system monitors differential inhibition of microbial growth by a panel of 18 dyes, antibiotics, and other chemicals. The effects of these inhibitors can be measured after 3 to 6 hr of incubation so that rapid identifications are feasible with this system.

The Autobac® Series II system consists of a disposable 19 chamber cuvette and the Autobac® instruments, including the disk dispenser, light-scattering photometer, incubator-shaker, and data processing system (IBM® PC computer [which replaces the previous Apple® IIe computer], a 20-megabyte Winchester hard disk, Okidata® dot matrix printer, and color graphic monitor). Identification of gram-negative bacilli is the primary application of this system, although a presumptive identification of enterococci and *Staphylococcus saprophyticus* can be obtained and an identification panel of tests for yeasts is reported to be under development.

Identification of gram-negative bacilli with the Autobac® system requires knowledge of six preliminary test results: indole and oxidase reactivity, swarming growth on sheep blood agar, and presence or absence of growth, lactose fermentation, and precipitation of bile salts on MacConkey agar. The inhibitory test results cannot be analyzed by the computer system unless these preliminary tests are first entered into the system. Thus, the identification results will be delayed if mixed bacterial growth on the initial isolation media prevents accurate determination of the preliminary tests. The frequency of this problem, however, has not been reported in the published evaluations of this system.

To inoculate the Autobac® cuvette, isolated colonies are suspended in phosphate-buffered saline and adjusted to a standardized turbidity using the Autobac® photometer. The same adjusted inoculum can be used for identification and susceptibility testing. The inoculum is diluted 1:10 in the Autobac® low-thymidine eugonic broth and 30 mℓ is inoculated into the 19-chamber Autobac® cuvette containing 18 different identification disks (the remaining chamber is a growth control chamber). The cuvette is then incubated at 36°C with rotary agitation at 220 rpm. Bacterial growth is measured by the Autobac® photometer after 3 hr and the readings are automatically entered into the computer system. If growth in the control chamber is insufficient, then additional readings can be performed for up to 6 hr. The identification of the organism is determined using quadratic discriminant analysis of the tests and the relative probability of the first and second choice organisms are printed on the report form. A low probability of identification (i.e., <0.80) indicates that supplementary conventional tests are required.

For the initial development of this identification system, 31 antimicrobial agents were analyzed, of which 18 were selected for further testing.[37] The 18 antimicrobials were then tested against 481 bacterial isolates, of which 97.3% were identified correctly at the genus or species level with the majority of identification discrepancies encountered with *Citrobacter* and *Enterobacter* isolates. Because the validity of this identification approach may be compromised by the development of resistance to commonly used antimicrobial agents, the use of nonantibiotic chemical compounds was examined.[8,38] More than 600 compounds were identified, of which 4 compounds (brilliant green, chlorhexidine diacetate, D-cycloserine, and gallacetophenone) were selected initially.[8] Sielaff and associates[38] later modified this system to include the 18 compounds that are presently used in the Autobac® system (Table 5).

The Autobac® identification system database was developed for 30 different groups of gram-negative bacilli including the most common isolates of the family Enterobac-

Table 5
SELECTIVE AGENTS USED IN THE AUTOBAC IDENTIFICATION SYSTEM

Agent	Disk mass (μg)
Acriflavine	30
Brilliant green	3
Cobalt chloride	375
Cycloserine	78
Cycloserine	240
3,5-Dibromosalicyclic acid	750
Dodecylamine hydrochloride	18.7
Floxuridine	36
Malachite green	3
Methylene blue	255
Omadine disulfide	5.5
Sodium azide	75
Thallous acetate	150
Carbenicillin	40
Cephalothin	13.5
Colistin	13
Kanamycin	5.4
Novobiocin	48

teriaceae and other gram-negative bacilli. The non-Enterobacteriaceae that can be identified with this system are six species or groups in the genus *Pseudomonas, Acinetobacter calcoaceticus,* and the genera *Aeromonas, Alcaligenes, Flavobacterium,* and *Moraxella. Klebsiella* species other than *K. pneumoniae, Providencia, Salmonella, Serratia,* and *Shigella* can only be identified at the genus level with this system.

There have been 6 studies published that tested almost 10,000 organisms in the present configuration of the Autobac® identification system (Table 6). The overall accuracy of identification in these studies varied from 87.7 to 94.8%, with commonly isolated organisms such as *Pseudomonas aeruginosa, Escherichia coli, K. pneumoniae,* and *Proteus mirabilis* identified accurately. The isolates that posed the most frequent problems for the system were *Alcaligenes, Pseudomonas cepacia, Citrobacter freundii, Enterobacter agglomerans, E. cloacae, Klebsiella* species other than *K. pneumoniae, Proteus vulgaris, Providencia, Salmonella, Shigella,* and *Yersinia.* Test specificity, that is, the measure of confidence that can be given to a particular identification, was poor with certain identifications. As an example, although most studies reported isolates of *Edwardsiella tarda* were identified accurately, some strains of *Escherichia coli* were misidentified as *Edwardsiella tarda.* Thus, the identification of an isolate as *E. tarda* in the system was frequently incorrect. Other identifications with a low test specificity included *K. pneumoniae* (*K. pneumoniae* and *K. oxytoca* were not differentiated by the system), *Enterobacter agglomerans, E. aerogenes, E. cloacae, Citrobacter freundii, Yersinia enterocolitica, Providencia, Moraxella,* and *Pseudomonas* species.[7,28,40] Strains of organisms not included in the database (e.g., *Achromobacter, Bordetella, Eikenella, Pasteurella,* and *Vibrio*) were also frequently misidentified as *Moraxella, Pseudomonas* species, *Flavobacterium,* and members of the Enterobacteriaceae.[7,10,12] Although the manufacturer can rightfully claim these organisms should not be tested in the Autobac® system, unless the system assigns a low probability value to the identification of these isolates, misidentifications will be unavoidable.

Table 6
SUMMARY OF TEST ACCURACY OF AUTOBAC IDENTIFICATION SYSTEM

Number of organisms tested	Percentage of organisms with Correct identification	$p < 0.8$[a]	Ref.
3726	88.4	—[b]	38
2889	87.7	8.5	7
290	90.5	4.9	12
1404	92.7	—[b]	28
1076	90.3	7.1	40
403	94.8	8.9	10

[a] Probability of identification is <0.8; identifications at this level are equivocal and must be confirmed by additional conventional tests.
[b] Number of equivocal identifications not stated.

The accuracy and specificity problems with the Autobac® system are in large part due to the limited database that is used currently. For example, misidentification problems with *K. oxytoca* could be corrected by incorporating the spot indole test result in the database for this organism. Another example of the limitations of the database can be demonstrated by examining the identification accuracy that was reported from different testing institutions for specific groups of organisms. Kelly and associates[28] reported in their collaborative study that all four laboratories had testing problems (e.g., an identification accuracy of less than 90%) with only 5 of the 12 species. The test inaccuracies with the other seven groups of organisms were primarily observed in only one or two test laboratories.

Test reproducibility with the Autobac® identification system compared favorably with conventional reference methods, with intralaboratory reproducibility between 85 and 95% and interlaboratory reproducibility between 82 and 95%.[7,28,40] Sielaff and associates[38] reported that the identification accuracy for 4 control organisms tested during a 9-month period was between 97 and 100%.

Only limited published data about the rapidity of identification with the Autobac® system are available. Stevens and associates[40] reported that 93% of the isolates that they tested were identified within 5 hr, with 60% of the organisms identified during the first 3.5 hr of incubation. This study did not indicate if certain groups of organisms (e.g., nonfermenters) required a longer period of incubation before identification was available. Between 5 to 9% of tested isolates have a low relative probability of identification value (e.g., <0.80; Table 6) and, thus, conventional biochemical tests must be performed before an identification is available.[7,10,12,40]

D. COBAS Bact

The COBAS Bact was developed by Roche Diagnostics (Basel, Switzerland) for bacterial identification and antimicrobial susceptibility testing. The system consists of disposable polystyrene rotors and the COBAS Bact instrument with a centrifugal filling station, incubator, photometer, thermomatrix printer, keyboard, and display window for results. An antimicrobial disk loading device is used for susceptibility tests. The COBAS Bact tests are highly automated. The user standardizes and pipettes the test inoculum into the COBAS rotor, whereas all subsequent steps are performed in the instrument. The inoculum (200 $\mu\ell$) is dispensed by low-speed centrifugation into the

Table 7
COMPARISON OF RAPID AUTOMATED IDENTIFICATION SYSTEMS

	Vitek® system	Abbott systems	Autobac system
Identification capabilities (species)			
Gram-negative bacilli	70	39	41
Gram-positive bacteria	45	—	—
Yeast	27	34	—
Biochemical tests (number)			
Gram-negative bacilli	29	20	18
Gram-positive bacteria	29	—	—
Yeast	26	20	—
Inoculum size			
Gram-negative bacilli	McFarland 1—2 in 2 m*l*	McFarland 0.5 in 5 m*l*	McFarland 0.5 in 2 m*l*
Gram-positive bacteria	McFarland 0.5 in 2 m*l*		—
Yeast	McFarland 2 in 2 m*l*	McFarland 0.5 in 5 m*l*	—
Inoculum preparation	Manual	Manual	Manual
Incubation	Bacteria: on-line	Off-line	Off-line
	Yeast: off-line	Off-line	—
Incubation time (hr)			
Gram-negative bacilli	4—18	4—5	3—6
Gram-positive bacteria	4	—	—
Yeast	24	24	—
Additional tests required before interpretation of results	Yes (1)	Yes (1)	Yes (6)
Addition of reagents required	No	Yes (indole)	No

rotor chambers containing dried biochemical substrates. Afterwards, the rotor is automatically transported into the incubation chamber for incubation at a predetermined temperature for 4 to 5 hr. The rotor is then automatically transferred to the photometer where each test reaction is measured at four wavelengths (430, 546, 564, and 605 nm). After the test reactions are compared with the established database, the bacterial identification is printed.

At the present time, only the prototype of the identification system has been evaluated. This preliminary analysis demonstrated that approximately 90% of common Enterobacteriaceae isolates were identified correctly. However, before this automated, versatile instrument can be used in the clinical laboratory, additional work must be performed to analyze the selection of biochemical tests used for identification and to expand the database for fermentative and nonfermentative gram-negative bacilli.

III. CONCLUSIONS

Three major automated systems (Vitek® system, the Avantage system, and the Autobac® system) are currently available for the rapid identification of microorganisms (Table 7), with a fourth system (the COBAS Bact) to be released soon for this function. All three systems can be used to identify common isolates of Enterobacteriaceae and other gram-negative bacilli, although the Vitek® system has the most extensive database. The identification capabilities with the Autobac® system are particularly limited with many organisms identified only at the genus level. The Vitek® system is the only system that can be used for identification of gram-positive bacteria; the Vitek® and Avantage systems, but not the Autobac® system, can be used to identify yeasts. Each system can identify most common gram-negative bacilli accurately. However, less common isolates of Enterobacteriaceae and non-Enterobacteriaceae pose a challenge to the systems. For example, all systems had problems identifying some strains of

Enterobacter, Serratia, Citrobacter, Klebsiella species other than *K. pneumoniae*, and *Pseudomonas* species other than *P. aeruginosa*. Additionally, the specificity of uncommon identifications was poor with all systems. Thus, if an uncommon identification is reported, the identification should be confirmed. Use of the Vitek® system to identify gram-positive cocci and bacilli has limited utility. Although *Listeria, Aerococcus, Streptococcus pneumoniae*, common β-hemolytic streptococci, *Staphylococcus aureus, S. epidermidis*, and *S. saprophyticus* can be reliably identified with the Vitek® system, other species of coagulase-negative staphylococci and the viridans group of streptococci are not identified. Thus, the system is unable to identify the isolates that cause problems for other commercial and conventional identification systems. The Vitek® system can be used to identify most common isolates of yeast (albeit many identifications are reported with a low probability); however, the Abbott systems cannot be recommended at this time for this purpose.

The biochemical tests offered in the three systems are extensive, with traditional tests incorporated in the Vitek® and Avantage systems and a novel series of reagents selected to inhibit microbial growth in the Autobac® system. The design of the Avantage and Autobac cuvettes limits the number of biochemical tests that can be performed to only 20 and 18 tests, respectively, whereas as many as 30 tests can be used in the Vitek® system.

The inoculum must be prepared manually with all three systems, which has been noted to cause problems with yeast identifications with the Avantage system, and potentially can result in incorrect bacterial and yeast identifications with the other systems. To support this fact, most investigators have reported that retesting an isolate with aberrant biochemical reactions frequently results in a correct identification. No investigators have reported the frequency in which an adequate number of colonies required to prepare the test inoculum have been obtained. This may be a particular problem with the Vitek® system for which the heaviest inoculum is required. All three systems require the performance of additional tests before the test results can be interpreted by the instrument. For example, an oxidase test is required with the Vitek® and Avantage systems, and six additional tests are required with the Autobac system. This requirement can delay the identification if sufficient isolated colonies are not available for the preliminary tests.

Bacterial identification tests are incubated in the Vitek® system, whereas the other bacterial tests and all yeast identification tests are incubated off-line. The advantage of the on-line incubation is that the identification tests are completely automated after the inoculated tests are prepared, whereas the only aspect that is automated with the other tests is the interpretation of the test results. The Avantage and Autobac® systems can be used to identify gram-negative bacilli within 4 to 6 hr. However, the Vitek® system requires as long as 18 hr, a time significantly longer than that required with many commercially produced manual systems. It is important to note that many common Enterobacteriaceae can be identified with the Vitek® system in as few as 4 hr. Identification of gram-positive bacteria with the Vitek® system is performed in 4 hr and both the Vitek® and Avantage systems require 24 hr of incubation to identify yeasts.

Thus, these automated identification systems offer rapid identification of common gram-negative bacilli. Unfortunately, the Vitek® system, which offers the most extensive database and highest degree of automation, is also the slowest system. One would anticipate that with a different selection of biochemical tests this problem could be circumvented. Many of the problems observed with the other automated systems could be rectified by expansions of the database. Although this has been done in some instances, data is not currently available that can be used to evaluate these improvements.

REFERENCES

1. Almeida, R. J. and Jorgensen, J. H., Evaluation of the AutoMicrobic system gram-positive identification card for species identification of coagulase-negative staphylococci, *J. Clin. Microbiol.*, 18, 438, 1983.
2. Appelbaum, P. C., Jacobs, M. R., Heald, J. I., Palko, W. M., Duffett, A., Crist, R., and Naugle, P. A., Comparative evaluation of the API 20S system and the AutoMicrobic system gram-positive identification card for species identification of streptococci, *J. Clin. Microbiol.*, 19, 164, 1984.
3. Barry, A. L. and Badal, R. E., Identification of Enterobacteriaceae by the AutoMicrobic system: Enterobacteriaceae biochemical cards versus Enterobacteriaceae-plus biochemical cards, *J. Clin. Microbiol.*, 15, 575, 1982.
4. Barry, A. L. and Badal, R. E., Reliability of early identifications obtained with Enterobacteriaceae-plus biochemical cards in the AutoMicrobic system, *J. Clin. Microbiol.*, 16, 257, 1982.
5. Barry, A. L., Gavan, T. L., Badal, R. E., and Telenson, M. J., Sensitivity, specificity, and reproducibility of the AutoMicrobic system (with the Enterobacteriaceae-plus biochemical card) for identifying clinical isolates of gram-negative bacilli, *J. Clin. Microbiol.*, 15, 582, 1982.
6. Barry, A. L., Gavan, T. L., Badal, R. E., and Telenson, M. J., Direct comparison of two mechanized systems for identification of gram-negative bacilli, *Am. J. Clin. Pathol.*, 78, 462, 1982.
7. Barry, A. L., Gavan, T. L., Smith, P. B., Matsen, J. M., Morello, J. A., and Sielaff, B. H., Accuracy and precision of the Autobac system for rapid identification of gram-negative bacilli: a collaborative evaluation, *J. Clin. Microbiol.*, 15, 1111, 1982.
8. Buck, G. E., Sielaff, B. H., Boshard, R., and Matsen, J. M., Automated, rapid identification of bacteria by pattern analysis of growth inhibition profiles obtained with Autobac 1, *J. Clin. Microbiol.*, 6, 46, 1977.
9. Burdash, N. M., Teti, G., West, M. E., Bannister, E. R., and Manos, J. P., Evaluation of an automated, computerized system (AutoMicrobic system) for Enterobacteriaceae identification, *J. Clin. Microbiol.*, 13, 331, 1981.
10. Burdash, N. M., Welborn, A. L., Teti, G., Bannister, E. R., and Manos, J. P., Identification of gram-negative bacilli using the Autobac IDX, *Diagn. Microbiol. Infect. Dis.*, 3, 59, 1985.
11. Cooper, B. H., Prowant, S., Alexander, B., and Brunson, D. H., Collaborative evaluation of the Abbott yeast identification system, *J. Clin. Microbiol.*, 19, 853, 1984.
12. Costigan, W. J. and Hollick, G. E., Use of the Autobac IDX system for rapid identification of Enterobacteriaceae and nonfermentative gram-negative bacilli, *J. Clin. Microbiol.*, 19, 301, 1984.
13. Davis, J. R., Stager, C. E., Wende, R. D., and Hussain Qadri, S. M., Clinical laboratory evaluation of the AutoMicrobic system Enterobacteriaceae biochemical card, *J. Clin. Microbiol.*, 14, 370, 1981.
14. Dipersio, J. R., Dyke, J. W., and Vannest, R. D., Evaluation of the updated MS-2 bacterial identification system in comparison with the API 20E system, *J. Clin. Microbiol.*, 18, 128, 1983.
15. Facklam, R., Bosley, G. S., Rhoden, D., Franklin, A. R., Weaver, N., and Schulman, R., Comparative evaluation of the API 20S and AutoMicrobic Gram-positive identification systems for non-beta-hemolytic streptococci and aerococci, *J. Clin. Microbiol.*, 21, 535, 1985.
16. Ferraro, M. J. B., Edelblut, M. A., and Kunz, L. J., Accurate automated identification of selected Enterobacteriaceae at four hours, *J. Clin. Microbiol.*, 13, 151, 1981.
17. Freeman, J. W., Rowland, R. W., Overman, S. B., and Goodman, N. L., Laboratory evaluation of the AutoMicrobic system for identification of Enterobacteriaceae, *J. Clin. Microbiol.*, 13, 895, 1981.
18. Goldstein, J., Guarneri, J. J., Della-Latta, P., and Scherer, J., Use of the AutoMicrobic and Enteric-Tek systems for identification of Enterobacteriaceae, *J. Clin. Microbiol.*, 15, 654, 1982.
19. Grasmick, A. E., Naito, N., and Bruckner, D. A., Clinical comparison of the AutoMicrobic system gram-positive identification card, API Staph-Ident, and conventional methods in the identification of coagulase-negative *Staphylococcus* spp., *J. Clin. Microbiol.*, 18, 1323, 1983.
20. Hasyn, J. J. and Buckley, H. R., Evaluation of the AutoMicrobic system for identification of yeasts, *J. Clin. Microbiol.*, 16, 901, 1982.
21. Hasyn, J. J., Cundy, K. R., Dietz, C. C., and Wong, W., Clinical laboratory evaluation of the AutoMicrobic system for rapid identification of Enterobacteriaceae, *J. Clin. Microbiol.*, 13, 491, 1981.
22. Heinz, T., Pezzlo, M., and Thrupp, L., Clinical and laboratory efficiency of rapid screening urine cultures of nephelometry, *Am. J. Clin. Pathol.*, 77, 305, 1982.
23. Isenberg, H. D., Gavan, T. L., Smith, P. B., Sonnenwirth, A., Taylor, W., Martin, W. J., Rhoden, D., and Balows, A., Collaborative investigation of the AutoMicrobic system Enterobacteriaceae biochemical card, *J. Clin. Microbiol.*, 11, 694, 1980.

24. Jorgensen, J. H., Dyke, J. W., Helgeson, N. G. P., Cooper, B. H., Redding, J. S., Crawford, S. A., Andruszewski, M. T., and Prowant, S. A., Collaborative evaluation of the Abbott Avantage system for identification of frequently isolated nonfermentative or oxidase-positive gram-negative bacilli, *J. Clin. Microbiol.*, 20, 899, 1984.
25. Jorgensen, J. H., Johnson, J. E., Alexander, G. A., Paxson, R., and Alderson, G. L., Comparison of automated and rapid manual methods for the same-day identification of Enterobacteriaceae, *Am. J. Clin. Pathol.*, 79, 683, 1983.
26. Johnson, J. E. and Brinkley, A. W., Comparison of the AutoMicrobic system and a conventional tube system for identification of nonfermentative and oxidase-positive gram-negative bacilli, *J. Clin. Microbiol.*, 15, 25, 1982.
27. Kelly, M. T. and Latimer, J. M., Comparison of the AutoMicrobic system with API, Enterotube, Micro-ID, Micro-Media systems, and conventional methods for identification of Enterobacteriaceae, *J. Clin. Microbiol.*, 112, 659, 1980.
28. Kelly, M. T., Matsen, J. M., Morello, J. A., Smith, P. B., and Tilton, R. C., Collaborative Clinical Evaluation of the Autobac IDX System for Identification of gram-negative bacilli, *J. Clin. Microbiol.*, 19, 529, 1984.
29. Kiehn, T. E., Edwards, F. F., Tom, D., Lieberman, G., Bernard, E. M., and Armstrong, D., Evaluation of the Quantum II yeast identification system, *J. Clin. Microbiol.*, 22, 216, 1985.
30. Land, G., Stotler, R., Land, K., and Staneck, J., Update and evaluation of the AutoMicrobic yeast identification system, *J. Clin. Microbiol.*, 20, 649, 1984.
31. McCracken, A. W., Martin, W. J., McCarthy, R., Schwab, D. A., Cooper, B. H., Helgeson, N. G. P., Prowant, S., and Robson, J., Evaluation of the MS-2 system for rapid identification of Enterobacteriaceae, *J. Clin. Microbiol.*, 12, 684, 1980.
32. Murray, P. R., Gauthier, A., and Niles, A., Evaluation of the Quantum II and Rapid E identification systems, *J. Clin. Microbiol.*, 20, 509, 1984.
33. Oblack, D. L., Rhodes, J. C., and Martin, W. J., Clinical evaluation of the AutoMicrobic system yeast biochemical card for rapid identification of medically important yeasts, *J. Clin. Microbiol.*, 13, 351, 1981.
34. Pezzlo, M. T., Tan, G. L., Peterson, E. M., and de la Maza, L. M., Screening of urine cultures by three automated systems, *J. Clin. Microbiol.*, 15, 468, 1982.
35. Ruoff, K. L., Ferraro, M. J., Jerz, M. E., and Kissling, J., Automated identification of gram-positive bacteria, *J. Clin. Microbiol.*, 16, 1091, 1982.
36. Salkin, I. F., Schadow, K. H., Bankaitis, L. A., McGinnis, M. R., and Kemna, M. E., Evaluation of Abbott Quantum II yeast identification system, *J. Clin. Microbiol.*, 22, 442, 1985.
37. Sielaff, B. H., Johnson, E. A., and Matsen, J. M., Computer-assisted bacterial identification utilizing antimicrobial susceptibility profiles generated by Autobac 1, *J. Clin. Microbiol.*, 3, 105, 1976.
38. Sielaff, B. H., Matsen, J. M., and McKie, J. E., Novel approach to bacterial identification that uses the Autobac system, *J. Clin. Microbiol.*, 15, 1103, 1982.
39. Smith, S. M., Cundy, K. R., Gilardi, G. L., and Wong, W., Evaluation of the AutoMicrobic system for identification of glucose-nonfermenting gram-negative rods, *J. Clin. Microbiol.*, 15, 302, 1982.
40. Stevens, M., Feltham, R. K. A., Schneider, F., Grasmick, C., Schaak, F., and Roos, P., A collaborative evaluation of a rapid automated bacterial identification system: the Autobac IDX, *Eur. J. Clin. Microbiol.*, 3, 419, 1984.
41. Sylvester, M. K. and Washington, J. A., Evaluation of the Quantum II microbiology system for bacterial identification, *J. Clin. Microbiol.*, 20, 1196, 1984.
42. Woolfrey, B. F., Lally, R. T., Ederer, M. N., and Quall, C. O., Evaluation of the AutoMicrobic system for identification and susceptibility testing of gram-negative bacilli, *J. Clin. Microbiol.*, 20, 1053, 1984.

Instrument Methods in Current Use for Performing Antimicrobial Susceptibility Tests

Chapter 7

INSTRUMENT SYSTEMS WHICH UTILIZE A CONVENTIONAL INCUBATION PERIOD

Richard L. Cohen

TABLE OF CONTENTS

I. INTRODUCTION

Instrumentation in clinical microbiology has lagged far behind the level of automation now present in chemistry and hematology, particularly because microbiology analyses are complex and lend themselves poorly to automation, and because clinical microbiologists historically have been skeptical of unconventional methods. During the past 10 to 15 years these factors have been overcome, to some degree, by the development of numerous mechanized and somewhat automated instruments, not all of which will survive past the temporary infatuation of the microbiologist.

This chapter will examine those instrument systems that are designed for overnight determinations of antimicrobial susceptibilities. Basically, these systems represent mechanizations of preexisting manual systems, and, with one exception, utilize microdilution techniques. The single exception is a system that reports minimal inhibitory concentrations (MICs) that are extrapolated from conventional disk diffusion zone sizes. In the discussion to follow, the genealogy of the instrument systems will be traced, their individual features will be presented, and their relative advantages and disadvantages will be compared. In addition, the relative merits of the microdilution technique itself will be discussed; as selection of an instrument system first requires a decision as to the suitability of the basic technique.

Dilution of antimicrobial agents in broth is widely accepted as an accurate and reliable method for in vitro determination of antimicrobial susceptibilities. Unfortunately, the standardized broth macrodilution procedures[1,2] are too laborious and expensive to be used routinely. Microdilution techniques utilizing small plastic trays containing miniature wells were developed originally for serologic procedures, but in 1966, Maymont and Wentz[3] described their use for MIC testing. The original trays contained 96 wells arranged in 8 rows of 12. Basically, the procedure involves placing 0.05 m*ℓ* of broth into each row, making serial twofold dilutions using metal loops calibrated to hold 0.05 m*ℓ*, inoculating each well with 0.05 m*ℓ* of a standardized inoculum using a calibrated dropper, incubating the tray overnight, and reading results using a specially designed mirrored reader. Since its first description, the technique has undergone changes as described below, and it has been adopted for routine use by increasing numbers of clinical microbiology laboratories.[4,5]

Developments in microdilution susceptibility testing have been extensive and have taken two distinct paths. These are mechanization of intramural laboratory preparation of trays, and development of commercial systems incorporating premade trays and special inoculation, reading, and data mangement devices.

As the needs and capabilities of clinical microbiology laboratories vary greatly, selection of the most appropriate system for a given laboratory requires a thorough understanding of the individual features of each instrument system. Microdilution trays themselves are all quite similar, whether prepared in the laboratory or purchased commercially; however, ancillary features of the two approaches are different enough to warrant separate discussions of each. Before the various instrument systems can be discussed, it is necessary to establish the performance adequacy of the microdilution technique itself.

II. MICRODILUTION SUSCEPTIBILITY TESTING

A. General Considerations

Marymont and Wentz[3] first described the microdilution procedure, and, comparing it with the macrodilution standard, obtained 95% agreement within 2 twofold dilutions and 69% within 1 twofold dilution. Harwick, Weiss and Fekety[6] obtained 92.5%

agreement within 2 twofold dilutions and MacLowry and Marsh[7] demonstrated that 68% of the microdilution results produce the same endpoints as the macrodilution procedure, and that an additional 19% varied by only 1 dilution. Chitwood[8] and Gavan and Town[9] also demonstrated excellent correlation, especially after the technologists gained experience with the technique. In addition, numerous investigators demonstrated the technique to be extremely reproducible.[6,7,9-19]

In order to standardize the microdilution procedure, the effects of numerous variables have been studied. These include media[1,20] and its supplements,[21,22] inoculum size,[1,20] physiologic condition of the inoculum,[16,23,24] incubation time,[20] stability during storage,[10,25,26] evaporation,[10,20] dilution error,[27] and contamination.[6,20] The medium of choice was determined to be Mueller-Hinton broth; because it supports growth of most pathogens, displays batch to batch reproducibility, is low in inhibitory substances and has extensive data supporting its use. Supplementation with calcium and magnesium is recommended to improve reliability of results of *Pseudomonas aeruginosa* susceptibilities to the aminoglycosides,[2,21] and with NaCl to improve testing of penicillinase-resistant penicillins.[22] The recommended inoculum is 5×10^5 colony forming units/mℓ,[2] and current evidence indicates that it can be prepared directly from an overnight agar culture.[16,23,24] Overnight incubation for 16 to 20 hr at 35°C is recommended,[2] but a promising procedure has been developed by Gibco Laboratories, Inc. (Lawrence, Mass.) that utilizes measurement of fluorogenic substrates to allow reading in 5 hr. Frozen trays can be stored for a minimum of 6 weeks at −20°C or −70°C,[10,25,26] and dried trays can be stored at room temperature for at least 1 year.

From these combined experiences, it seems clear that the microdilution technique is highly acceptable for routine antimicrobial testing. Reservations concerning the method are not greater than those concerning any other technique, and, in a review of changing trends in susceptibility testing, Jones reported that greater than 25% of clinical microbiology laboratories in the U.S. currently utilize microdilution as their routine procedure.[5] With this premise established, the next step is to examine specific salient features of the various currently available systems.

B. Laboratory Preparation of Microdilution Trays

The original techniques for microdilution MIC tests utilized 96-well disposable plastic microdilution trays, calibrated droppers to add diluent and inoculum, and calibrated loops to make serial twofold dilutions.[3,6,8,9]

Even using these manual techniques, use of microdilution trays did offer improved ease and economy of MIC testing. Unfortunately, manual preparation was still very laborious. To decrease the labor of tray preparation, automatic serial diluters were fabricated[7,27] or purchased from manufacturers who designed them for use in serologic procedures.[20-25] These devices incorporated a set of microdiluters that simultaneously mixed the contents of each well in a row, then transfered a calibrated aliquot to the wells of the next row. This process was repeated and the diluters advanced automatically until all of the twofold dilutions were made. To facilitate preparation further, machines were developed that automatically added diluent.[7,27] Use of these machines did reduce labor and time, but production of adequately large numbers of trays was still too time consuming to be practical for most routine laboratories.

In addition to the problems associated with preparation of trays, the relevance of using serial dilutions was questioned. Serial twofold dilutions for MIC testing became established as the accepted format for the tube macrodilution procedure due to the relative ease of preparation and their long-standing use in microbiology and serology. However, as more became known about body site distribution and pharmacology of the antimicrobics, it became apparent that concentrations other than those obtained by serial dilutions should be tested.[15,28]

To facilitate preparation of large batches of trays, instruments were developed that had individual reservoirs for each concentration and were capable of rapidly dispensing all concentrations simultaneously. In addition to allowing large batch production, they also allowed free selection of concentrations to be tested, modification of the growth media in specific wells, and addition of biochemical substrates for bacterial identification.

The MIC®-2000 system (Dynatech Laboratories, Inc., Alexandria, Va.), introduced in 1975, was the first such system marketed commercially. Its use has been evaluated[29-31] and proven to be accurate, reproducible, relatively simple and reliable. Its use has been so widely accepted, it is used to prepare reference trays for evaluations of commercial trays[10,12-14,17,32,33] and for studies to evaluate variables associated with MIC testing.[15,34]

The MIC®-2000 system consists of a semiautomated 96-channel dispenser, a semiautomated tray inoculator, a sealing tape dispenser, and a reader. Individually prepared antimicrobic dilutions in 18 × 150 mm reservoir test tubes are placed in a rack that is inserted into the dispenser; 96 flexible tubes contained in a manifold are inserted into the reservoir tubes, and the chamber is closed and pressurized. Individual empty trays are then inserted into the instrument, and antimicrobics are delivered into all the wells simultaneously in volumes of 0.05 or 0.1 mℓ. Although preparation of the individual dilutions is laborious, tray dispensing is simple and extremely rapid. After filling, the trays may be sealed in plastic bags and frozen at −20°C or less. A large batch of trays usually requires a full day of labor by a competent individual. If 0.1 mℓ is dispensed, approximately 200 trays can be filled with one reagent preparation, but this can be expanded by preparing the antimicrobics in larger volumes and splitting them to be dispensed in separate runs.

Since the introduction of the MIC®-2000, other similar systems have been developed. Bellco Glass Inc. (Vineland, N.J.) distributes a set of instruments manufactured by Sandy Spring Instrument Co. (Denver, Colo.) that operate in a manner similar to the MIC®-2000. Three individual dispensers, having capacities of 20 mℓ, 100 mℓ, and 200 mℓ are available, and as many as 1500 trays can be dispensed per hour. In addition, Dynatech now produces systems with significantly larger capacities.

Carrying laboratory preparation and testing one step further, Turner et al.[35] and Courcol et al.[36] interfaced a microcomputer with a commercially available photometer (Dynatech® MR-580 Microelisa Autoreader, Dynatech Laboratories Ltd., Billingshurst, England) that was designed for reading microelisa trays. Turner et al.[35] obtained exact agreement with visual reading in 81.2% of the tests and agreement within one dilution in 98.6%. Courcol et al.[36] reported 98.8% agreement, but did not stipulate whether this was exact or within a specific range of dilutions.

C. Commercial Microdilution Systems

1. General Considerations

With the development of the semiautomated instruments for tray preparation, MIC testing gained wide acceptance. However, tray preparation was still not attractive to most laboratories, and it was not until the first commercially prepared trays became available that MIC testing gained wide acceptance in clinical laboratories. Since the market was first established by Pasco Laboratories, Inc. (Wheat Ridge, Colo.) numerous other systems have been introduced. At this time, at least six systems using microdilution trays with multiple antibiotics are available. In addition, Austin Biological Laboratories, Inc. (Austin, Tex.) and Analytab Products (Plainview, N.Y.) supply single antibiotic strips, but there is no instrumentation associated with their use, and they will not be discussed further. The multiple antimicrobic systems, listed alphabetically, include Micro-Media® (Beckman Instruments, Inc., Carlsbad, Calif.), Micro Scan® (American Micro Scan, Mahwah, N.J.), Pasco® (Pasco Laboratories, Inc., Wheat Ridge, Colo.), Sceptor® (BBL Microbiology Systems, Cockeysville, Md.), Sensititre® (Gibco Laboratories, Lawrence, Mass.), and UniScept (Analytab Products, Plainview, N.Y.). In addition, LANcraft, Inc. (Norcross, Ga.) has developed a system that includes a computer-based data mangement system and a manual reader that they claim can be modified for use with any commercial microdilution tray. When examining features of each system, one must keep in mind that all components of the system must function together satisfactorily. A superbly designed and functioning instrument system is worthless if the test trays themselves are designed poorly, and vice versa.

Selected features of each available system are given in Table 1. These should be used as background information only, as many of these systems will have changed by the publication of this book. Significance of the features and variables associated with them are discussed below, and these discussions may be used as a general guide in comparing systems.

2. Specific Features

a. Frozen vs. Dried Trays

The trays may be obtained in either frozen liquid or dried states. Micro-Media®, MicroScan®, and Pasco® trays are supplied in a frozen liquid state and thus require storage at a maximum temperature of about −20°C; whereas Sceptor®, Sensititre®, and UniScept are supplied in a dry state and can be stored at room temperature for extended periods. Each type has its advantages and disadvantages.

Frozen trays must not be allowed to thaw until used by the laboratory. This can present logistic problems, especially in more remote areas where supply channels are less than optimal. They also require utilization of a significant amount of freezer space, but this can be minimized by the judicious establishment of standing orders and frequent, small volume shipments. On the positive side, frozen panels are simpler and faster to inoculate by the laboratory. Hydration and inoculation of dried panels is more time consuming and requires use of single- or multi-tipped pipettes, or, for greatest convenience, the relatively expensive instruments sold by the respective manufacturers. These differences require careful consideration in light of local fiscal, personnel, and space considerations.

b. Number of Wells and Available Batteries

Selection of antimicrobics for testing 10 years ago was much simpler than now, since far fewer drugs were available. A plethora of new antimicrobics each with its own spectrum of activity now exists, and continued release of still newer drugs seems likely. Each laboratory is likely to have a somewhat different set of drugs which must be tested. Two potential solutions to the problem of which drugs to test have been developed by the manufacturers. One relates to production of large capacity panels that include relevant concentrations of essentially all FDA-approved antimicrobics, while the other relates to production of several different panels, each containing a different battery of antimicrobics. Each approach attempts to solve the problem without causing excessive confusion and expense.

The large capacity panels remove subjectivity from battery selection, and allow accumulation of more epidemiology markers and more data upon which future usage patterns can be established. Their use does increase cost somewhat, however, and it introduces the potential for reporting and subsequently using antimicrobics that may be inappropriate for the specific organism at the site of infection or in the particular patient. Systems utilizing large panels do allow tailoring of indivual reports to include

Table 1
FEATURES OF SPECIFIC COMMERCIAL MICRODILUTION PANELS

System	Physical state	No. of wells	Batteries available[a]	Customized panels
Micro-Media®	Frozen	150	Gram neg./UTI MIC/ID	No
		118	Gram neg. UTI MIC	
		118	Gram pos. MIC/ID	
		70	Anaerobic MIC	
		70	"Peer"[b] antibiotics	
Micro Scan®	Frozen	96	Gram neg. MIC/ID	No
		96	Gram neg. MIC	
		96	UTI MIC/ID	
		96	Gram pos. MIC/ID	
		96	Gram pos. MIC	
		150	MIC/Gram neg. ID	
Pasco®	Frozen	104	Gram neg. MID/ID	No
		104	UTI MIC/ID	
		104	Gram pos. MIC	
		104	Gram neg. MIC	
		104	UTI MIC	
		104	Resistance MIC	
Sceptor®	Dried	84	Gram neg. MIC/ID	No
		84	Gram neg. MIC	
		84	UTI MIC/ID	
		84	UTI MIC	
		84	Gram pos. MIC/ID	
		84	Gram pos. MIC	
		84	Anaerobic MIC/ID	
		84	Anerobic MIC	
Sensititre®	Dried	96	Gram pos. MIC	Yes
		96	Gram neg. MIC	
		96	UTI MIC	
		96	Gram pos. K-B	
		96	Gram neg. K-B	
		96	Gram neg. ID	
UniScept	Dried	120	Gram pos. and neg. MIC	Yes
		40	Gram pos. and neg. K-B	
		10	Single antimic. MIC	

[a] UTI = urinary tract infection; MIC = minimal inhibitory concentration; ID = identification; K-B = Kirby-Bauer breakpoint concentrations.
[b] "Peer" antibiotics are those included in a class, but having slightly different spectra.

only selected antimicrobics. Future computer programs may assist in tailoring the report to the organism name and susceptibility pattern, to the body site, and to relevant conditions of the patient, as well as allowing antimicrobics to be reported on demand.

The major advantage of the multiple batteries of smaller capacity panels is the lower cost. The majority of organisms isolated will only require testing by the first line battery, and a second panel can be set up only when needed. However, this will result in a time delay that may be unsatisfactory. As is evident from Table 1, both approaches are offered currently. To select the more satisfactory approach, each laboratory must analyze its individual needs and capabilities, susceptibility patterns of organisms isolated from its patient population, and, especially, antimicrobic usage by the physicians served.

c. Panel Customization

Another, perhaps more advantageous, possibility is the ability to purchase customized test panels. At the present time, only the Gibco Sensititre® system allows production of customized panels, although API has announced its intention to provide a similar service. Unfortunately, customization may be a problem in regard to FDA clearance, as well as the necessary customization of an instrument system's computer program. In addition, this requires close cooperation between the laboratory and clinical staff to determine which antimicrobics to test, as well as purchase of a large number of trays of the special configuration. One alternative may be the use of individual antimicrobic strips to supplement standard panels when testing of additional drugs is necessary.

d. Sources of Supply

Although not relating directly to instrumentation, sources of supply for the trays must be considered, especially for those systems that utilize frozen trays. Reliability of supply, cost of shipping, efficiency of the hospital's receiving and distribution system, and the manufacturer's guarantee must all be considered.

e. Cost

Finally, costs of trays and supporting instruments cannot be ignored. Since prices change constantly, it is not possible to compare them accurately in this chapter. Two facts can be stated, however. First, the market is highly competitive, and costs for the different systems must be assessed frequently. Second, competition among manufacturers can favorably benefit the laboratory by bringing about price reductions.

3. Instrumentation

Instrumentation supporting microdilution systems has developed in a series of overlapping steps, beginning with the simple light box readers and evolving to the present relatively high level of sophistication. Now there are semiautomated reconstitution/inoculation stations, highly refined manual or automated readers, and state-of-the-art computers supported by dynamic and highly sophisticated software. In addition, at least two fully automated systems are nearing completion.

Since various systems differ from one another in detailed features which are frequently undergoing change, no detailed discussion of the individual features will be attempted here. Instead, the discussion will be directed toward a simple description of the various instrument components available at this time, and a more detailed description of the possible distinguishing features that should be examined by a potential purchaser. Components of the various systems are given in Table 2.

a. Tray Set Up

Systems having frozen trays utilize disposable plastic inoculators that consist of an inoculum reservoir and a cover having a sufficient number of plastic prongs to deliver standardized inocula simultaneously to all wells. Use of these inoculators is simple and rapid. Conversely, systems using dried trays require a more complex method for reconstitution and inoculation. Although this can be accomplished manually using multitipped pipettes, it is much more efficient to employ an instrument. Sceptor®, Sensititre®, and UniScept all use instruments that simultaneously dispense the inoculum and test medium from a disposable reservoir into each well in a stepwise manner. Inoculation of dried trays currently is slower than inoculation of frozen trays, and the instruments can malfunction, resulting in loss of trays and time.

Table 2
INSTRUMENT COMPONENTS

mponent	Instrument system						
	Micro-Media®	Micro Scan®	Pasco®	Sceptor®	Sensititre®	UniScept	LANcraft[a]
nstitution/ on ler	No	No	No	Yes	Yes	Yes[b]	
	Yes	Yes	Yes	Yes	Yes	Yes	Yes
ted	No	Yes	No	No	Yes[c]	Yes	No
	TeleVideo 803H	IBM® PC XT	NCR® Worksaver 300	Digital® Professional 350	Digital® Professional 350	Digital® based plus IBM® PC and MacIntosh®	TeleVideo 803H 04 1605 H or TeleXT or IBM® PC XT
	Anadex Silent/ Scriber	Anadex Silent/Scriber	Okidata 92	Digital LA50-RA	Digital® LA50	Epson FX 80	Anadex Silent/Scriber

nent system only; utilizes panels from other manufacturers.

struments available: one inoculates in a stepwise fashion; the other inoculates all wells at once.

5-hr reading of MICs using fluorogenic substrates.

b. Reader

Manual and automated devices for reading and interpreting MIC tray results also are available. The manual readers consist of light boxes designed to enhance visualization of reactions and some method of result entry. Micro-Media®, Pasco®, Sensititre® and UniScept all have readers with LEDs that display below the tray and indicate which result has been chosen. This is an excellent and highly desirable feature that helps prevent entry error by the operator. The method of entry is different for each system. Micro-Media® utilizes a "joy stick" that the operator uses to move a cursor around the tray; Micro Scan® utilizes a sonic stylus; Pasco®, a light pen; Sceptor®, a bar with a touchpad that is moved over each row of wells; Sensititre®, a keyboard; and UniScept, a touchpad keyboard. Each method works effectively, and an experienced operator can read and record results rapidly.

For the laboratory that does not wish to invest in a full data mangement system, the Micro-Media®, Sceptor®, and UniScept sysems have less expensive, stand-alone reading capabilities. Micro-Media® and UniScept both can print MIC and identification results. The Micro-Media® reader currently requires no additional equipment, whereas that of UniScept requires a printer and interface with a computer having limited data management capabilities. Sceptor® provides an MIC report giving Kirby-Bauer breakpoints, but organism identification must be determined and entered manually.

Automated photometric readers are available from Micro Scan®, Sensititre®, and UniScept. The theoretical advantages of automated readers are reduced labor, greater speed, and reading consistency. Evaluations of the readers[37-39] confirm their adequacy, although performance is not better than that of manual interpretation.[40] The Sensititre® reader does possess a potential advantage in that fluorogenic substrates can be added to the inoculum, making it possible to determine results after only 5 hr of incubation. Automated readers add significantly to the cost of the system, and potential purchasers must carefully consider their benefits in light of the increased costs.

c. Data Management

Each of the systems offers a free-standing microcomputer that is fast, has high-capacity storage, and is user friendly. These include popular personal computers such as IBM® PC, MacIntosh® by Apple, and Digital® Equipment Co. models. In addition, the laboratory can utilize the computers' other capabilities, such as word processing, inventory control, work load calculations, etc. All of the systems now have multiuser, multitasking capabilities. In addition, API has chosen to develop a distributive processing network system that utilizes a microbiology data system (MDS) that can interface with any of the UniScept components, and has full logbook and epidemiology capabilities. IBM® PC and Apple MacIntosh® computers are also available, and these can stand alone or network with the MDS. This networking system offers greater versatility and allows continuity of function if one of the computers becomes inoperative.

Logbook, epidemiology, and other software capabilities are similar in all systems, but they are developing and changing rapidly and cannot be detailed fully here. Instead, general features of data management capabilities will be discussed below, and these can be used as guides for examining any of the systems.

4. Data Management Capabilities

The systems are being developed for total microbiology laboratory data handling, and to accomplish this, they should possess the features given below. In addition, they should require minimal clerical activity, be capable of interfacing with mainframe computers, and be capable of having many or most of their services customized by the laboratory.

a. Patient Reports

Patient reports can have a variety of formats, and the laboratory should be able to customize these. Reports should include relevant patient and specimen information, direct smear results, organism identification and relative numbers present, antimicrobic susceptibilities and interpretations, and free text or coded comments. Including additional tests, e.g., fungal and mycobacterial cultures, for a given specimen on the same report also may be desirable.

Organism names should be derived from the microdilution identification systems or entered manually in free text or code if they are derived by other means. Manual entry of names should be simple and fast, and should not necessitate undue disruption of work flow. The system's database should be extensive, including essentially all relevant organisms which can be tested in the manufacturer's trays, as well as other microorganisms that can be identified separately and entered into the computer manually. The database should be updated easily when new organism designations come into common use.

Antimicrobial susceptibility reports should include MIC results, as well as clinically useful interpretations. In addition, dosing guides and cost estimates that can be customized by the laboratory should be available. Several alternatives for MIC interpretation exist,[2,30,41] and until one becomes accepted universally, the laboratory should be able to select the most acceptable one.

Smear results and comments can be entered in free text or by codes established by the laboratory. Finally, the reports should be available at any stage of processing.

b. Data Retrieval

The laboratory should be able to retrieve data by several criteria, including patient name or identification number. If retrieval is by name, the system should be able to recognize various forms of the name as the same patient, as this will prevent loss of data because the name was entered differently. In addition, the laboratory should be able to retrieve data based on criteria it sets. For example, a laboratory may wish to recover results by organism name or biotype, by specific susceptibility patterns, as submitted by a specific physician or from a specific ward, etc. Also, retrieval by combinations of these criteria would be desirable, e.g., methicillin-resistant *Staphylococcus aureus* from patients treated by a specific physician.

c. Logbook Capabilities

Having the capability to function as a logbook can be a boon or burden. Manual entry of all patient and specimen information for all cultures can impose a prohibitive clerical burden; however, for the noncomputerized laboratory, the ability to generate a daily printed log listing all specimens and current results can reduce disruption of work flow arising from the need to answer physicians' requests for results. Logbooks can take several forms, including specimen logs with no results, logs containing only final results, and logs containing all results as they become available. They should group specimens by several criteria, including accession numbers and patient names.

d. Cumulative Reports

Cumulative antibiograms and epidemiology reports also may be prepared readily by the computer. Antibiograms can take several forms, including susceptibility percentages, percentages at each MIC level, etc. Epidemiology reports can also take several forms to include rate of isolation of specific organisms from specific wards or body sites. Some customization of report appearance and content should be provided. It may also be helpful for infection control purposes if the computer could "flag" results when numbers exceed laboratory established thresholds.

e. Other Capabilities

Storage and retrieval of all types of quality assurance data, as well as work load recording should be possible. Optional capabilities should include inventory control, materials ordering, and word processing.

f. Future Expansion

Because of the rapidity of developments, systems must be capable of being updated and upgraded with minimal cost and effort. Usually, software updates are provided at no cost by the manufacturers. This is not true for changes in hardware; therefore, one must seek assurance that existing hardware is likely to be able to handle future improvements.

III. MIC EXTRAPOLATION FROM DISK DIFFUSION ZONE SIZES

The recently introduced BIOGRAM (Giles Scientific, Inc., New York, N.Y.) translates disk diffusion zone sizes into MICs, using regression line analysis. The system includes an electronic caliper and light box for reading zone sizes and transmitting them to the preprogrammed computer that converts the results into MICs. A printed report is produced that includes calculated MICs, Kirby-Bauer interpretations, inhibitory quotients that are based on achievable serum, urine, bile, and cerebrospinal fluid concentrations[41] and cost factors. Using this system, MICs extrapolated from the zone sizes showed greater than 90% correlation with those obtained by reference microdilution procedures according to data provided by the manufacturer and D'Amato, et al.[42] Advantages of the system are said to be lower costs for supplies and equipment as compared with commercial microdilution systems, ability to select from 34 antimicrobics, more rapid addition of new antimicrobics, and the ability to read results for many organisms in 5 to 6 hours.

Potential disadvantages are some of the same as those inherent in the disk diffusion procedure. These include the inability to test slow-growing, capnophylic, and anaerobic bacteria[43] and the fact that zone sizes for some strains fall far from the regression line and can yield erroneous MIC extrapolations.

IV. SUMMARY

To select the most appropriate susceptibility testing system, one must consider carefully the laboratory's and physicians' needs. Some laboratories will find one of the microdilution methods highly desirable, while others may prefer traditional disk diffusion or automated systems, which provide rapid results.

The final method/instrument choice is not simple. Contacting laboratories that use a particular system can yield valuable information that cannot be obtained from other sources. Nuances of the system's function, its reliability during routine use, service support by the manufacturer, and other essential factors can be learned candidly in this manner. Finally, the instrument should be brought into the laboratory for an intramural evaluation. All potential users should be given the opportunity to work with it, and it should be decided whether the system possesses capabilities that are needed critically, and how it will fit the existing work flow. Only a thorough understanding of the laboratory's and hospital's own needs and capabilities, as well as of the absolute and relative capabilities of all the available systems will permit a logical, well-informed choice.

REFERENCES

1. Ericsson, H. M. and Sherris, J. C., Antibiotic sensitivity testing. Report of an international collaborative study, *Acta Pathol. Microbiol. Scand. Sect. B,* 217(Suppl.), 1, 1971.
2. National Committee for Clinical Laboratory Standards, *Tentative standard M7-T, Methods for dilution antimicrobial susceptibility tests for bacteria that grow aerobically,* National Committee for Clinical Laboratory Standards, Villanova, 1983.
3. Marymont, J. H. and Wentz, R. M., Serial dilution antibiotic sensitivity testing with the microtitrator system, *Am. J. Clin. Pathol.,* 45, 548, 1966.
4. Murray, P. R. and Jorgensen, J. H., Quantitative susceptibility test methods in major United States medical centers, *Antimicrob. Agents Chemother.,* 20, 66, 1981.
5. Jones, R. N., Antimicrobial susceptibility testing (AST): a review of changing trends, quality control guidelines, test accuracy, and recommendation for the testing of β-lactam drugs, *Diagn. Microbiol. Infect. Dis.,* 1, 1, 1983.
6. Harwick, H. H., Weiss, P., and Fekety, F. R., Jr., Application of microtitration techniques to bacteriostatic and bactericidal antibiotic susceptibility testing, *J. Lab. Clin. Med.,* 72, 511, 1968.
7. MacLowry, J. D. and Marsh, H. H., Semiautomatic microtechnique for serial dilution-antibiotic sensitivity testing in the clinical laboratory, *J. Lab. Clin. Med.,* 72, 685, 1968.
8. Chitwood, L. A., Tube dilution antimicrobial susceptibility testing: efficacy of a microtechnique applicable to diagnostic laboratories, *Appl. Microbiol.,* 17, 707, 1969.
9. Gavan, T. L. and Town, M. A., A microdilution method for antibiotic susceptibility testing: an evaluation, *Am. J. Clin. Pathol.,* 53, 880, 1970.
10. Barry, A. L., Jones, R. N., and Gavan, T. L., Evaluation of the Micro-Media system for quantitative antimicrobial drug susceptibility testing: a collaborative study, *Antimicrob. Agents Chemother.,* 13, 61, 1978.
11. Phillips, I., Warren, C., and Waterworth, P. M., Determination of antibiotic sensitivities by the Sensititre system, *J. Clin. Pathol.,* 31, 531, 1978.
12. Gavan, T. L., Jones, R. N., and Barry, A. L., Evaluation of the Sensititre system for quantitative antimicrobial drug susceptibility testing: a collaborative study, *Antimicrob. Agents Chemother.,* 17, 464, 1980.
13. Jones, R. N., Thornsberry, C., Barry, A. L., and Gavan, T. L., Evaluation of the Sceptor microdilution antibiotic susceptibility testing system: a collaborative study, *J. Clin. Microbiol.,* 13, 184, 1981.
14. Jones, R. N., Barry, A. L., Bigelow, J., Gavan, T. L., and Thornsberry, C., Evaluation of the MICUR system for quantitative antimicrobial susceptibility testing: a multiphasic comparison with reference methods, *J. Clin. Microbiol.,* 16, 153, 1982.
15. Woolfrey, B. F., Fox, J. M. K., Lally, R. T., and Quall, C. O., Broth microdilution testing of *Pseudomonas aeruginosa* and aminoglycosides: need for employing dilutions differing by small arithmetic increments, *J. Clin. Microbiol.,* 16, 663, 1982.
16. Barry, A. L., Badal, R. E., and Hawkinson, R. W., Influence of inoculum growth phase on microdilution susceptibility tests, *J. Clin. Microbiol.,* 18, 645, 1983.
17. Gerlach, E. H., Jones, R. N., and Barry, A. L., Collaborative evaluation of the microbial profile system for quantitative antimicrobial susceptibility testing, *J. Clin. Microbiol.,* 17, 436, 1983.
18. Isenberg, H. D., D'Amato, R. F., McKinley, G. A., Hochstein, L., and Sampson-Scherer, J., Collaborative evaluation of the UniScept quantitative antimicrobial susceptibility test, *J. Clin. Microbiol.,* 19, 733, 1984.
19. D'Amato, R. F., Isenberg, H. D., McKinldy, G. A., Hochstein, L., and Cleri, D. J., Collaborative evaluation of the UniScept qualitative antimicrobial susceptibility test, *J. Clin. Microbiol.,* 21, 293, 1985.
20. Tilton, R. C., Lieberman, L., and Gerlach, E. H., Microdilution antibiotic susceptibility test: examination of certain variables, *Appl. Microbiol.,* 26, 658, 1973.
21. Cherne, J. E. and Perterson, L. R., Microdilution aminoglycoside susceptibility testing of *Pseudomonas aeruginosa* and *Escherichia coli* with a cation-supplemented inoculum, *Antimicrob. Agents Chemother.,* 18, 210, 1980.

22. Thornsberry, C. and McDougal, L. K., Successful use of broth microdilution in susceptibility tests for methicillin-resistant (heteroresistant) staphylococci, *J. Clin. Microbiol.*, 18, 1084, 1983.
23. Baker, C. N., Thornsberry, C., and Hawkinson, R. W., Inoculum standardization in antimicrobial susceptibility testing. Evaluation of overnight agar cultures and the rapid inoculum standardization system, *J. Clin. Microbiol.*, 17, 450, 1983.
24. Wicks, J. H., Nelson, R. L., and Krejcarek, G. E., Rapid inoculum standardization system: a novel device for standardization of inocula in antimicrobial susceptibility testing, *J. Clin. Microbiol.*, 17, 1114, 1983.
25. Barry, A. L., Effinger, L. J., and Badal, R. E., Short-term storage of six penicillins and cephalothin in microdilution trays for antimicrobial susceptibility tests, *Antimicrob. Agents Chemother.*, 10, 83, 1976.
26. Kuglin Bailey, P. L. and McGuckin, M. B., Storage of microtiter plates for antimicrobial susceptibility tests, *Am. J. Med. Technol.*, 45, 517, 1979.
27. MacLowry, J. D., Jaqua, M. J., and Selepak, S. T., Detailed methodology and implementation of a semi-automated serial dilution microtechnique for antimicrobial susceptibility testing, *Appl. Microbiol.*, 20, 46, 1970.
28. Barry, A. L., Thornsberry, C., and Jones, R. N., Gentamicin and amikacin disk susceptibility tests with *Pseudomonas aeruginosa:* definition of minimal inhibitory concentration correlates for susceptible and resistant categories, *J. Clin. Microbiol.*, 13, 1000, 1981.
29. McMaster, P. R. B., Robertson, E. A., Witebsky, F. G., and MacLowry, J. D., Evaluation of a dispensing instrument (Dynetech MIC-2000) for preparing microtiter antibiotic plates and testing their potency during storage, *Antimicrob. Agents Chemother.*, 13, 842, 1978.
30. Witebsky, F. G., MacLowry, J. D., and French, S. S., Broth dilution minimum inhibitory concentrations: rationale for use of selected antimicrobial concentrations, *J. Clin. Microbiol.*, 9, 589, 1979.
31. Woolfrey, B. F., Ramadei, W. A., and Quall, C. O., Evaluation of a semi-automated micro-broth dilution system for determining minimum inhibitory concentrations of antimicrobics, *Am. J. Clin. Pathol.*, 73, 374, 1980.
32. Jones, R. N., Gavan, T. L., and Barry, A. L., Evaluation of the Sensititre microdilution antibiotic susceptibility system against recent clinical isolates: three laboratory collaborative study, *J. Clin. Microbiol.*, 11, 426, 1980.
33. Woolfrey, B. F., Lally, R. T., and Quall, C. O., Comparative evaluation of the Micro-Media system, Sceptor, and MIC-2000 microdilution methods for testing *Pseudomonas aeruginosa* against gentamicin, tobramycin, and amikacin, *J. Clin. Microbiol.*, 17, 312, 1983.
34. Woolfrey, B. F., Fox, J. M., and Quall, C. O., A comparison of minimum inhibitory concentration values determind by three antimicrobic dilution methods for *Pseudomonas aeruginosa*, *Am. J. Clin. Pathol.*, 75, 39, 1981.
35. Turner, A., Hawdey, P. M., and Pedler, S. J., Computer-assisted determination of antibiotic susceptibility in photometer-read microtiter plates, *J. Clin. Microbiol.*, 18, 996, 1983.
36. Courcol, R. J., Deleersnyder, H., Roussel-Delvallez, M., and Martin, G. R., Automated reading of a microtitre plate: preliminary evaluation in antimicrobial susceptibility tests and *Enterobacteriaceae* identification, *J. Clin. Pathol.*, 36, 341, 1983.
37. Ellner, P. D. and Myers, D. A., Preliminary evaluation of the autoSCAN-3, an instrument for automated reading and interpretation of microdilution trays: identification of aerobic gram-negative bacilli, *J. Clin. Microbiol.*, 14, 326, 1981.
38. DeGirolami, P. C., Eichelberger, K. A., Salfity, L. C., and Rizzo, M. F., Evaluation of the autoSCAN-3, a device for reading microdilution trays, *J. Clin. Microbiol.*, 18, 1292, 1983.
39. Woolfrey, B. F., Lally, R. T., and Quall, C. O., Evaluation of AutoSCAN-3 and Sceptor systems for Enterobacteriaceae identification, *J. Clin. Microbiol.*, 17, 807, 1983.
40. Barry, A. L. and Braun, L. E., Reader error in determining minimal inhibitory concentrations with microdilution susceptibility test panels, *J. Clin. Microbiol.*, 13, 228, 1981.
41. Ellner, P. D. and Neu, H. C., The inhibitory quotient. A method for interpreting minimum inhibitory concentration data, *J. Am. Med. Assoc.*, 246, 1575, 1981.
42. D'Amato, R. F., Hochstein, L., Vernaleo, J. R., and Cleri, D. J., Evaluation of the BIOGRAM antimicrobial susceptibility test system, *85th Annu. Meet. Am. Soc. Microbiol.*, American Society for Microbiology, Washington, D.C., 1985, 314.
43. National Committee for Clinical Laboratory Standards, *Approved Standard M2-A3. Performance Standards for Antimicrobial Disk Susceptibility Tests*, 3rd ed., National Committee for Clinical Laboratory Standards, Villanova, 1984.

Chapter 8

INSTRUMENT SYSTEMS WHICH PROVIDE RAPID (3- TO 6-HR) ANTIBIOTIC SUSCEPTIBILITY RESULTS

James H. Jorgensen

TABLE OF CONTENTS

I. INTRODUCTION

A laboratory may choose to perform automated antimicrobial susceptibility testing for one or more of the following reasons: (1) To provide test results more rapidly than can be accomplished by manual methods, (2) To reduce the amount of labor required for such tasks, or (3) To improve standardization of test results from day-to-day or between different laboratories. The use of an automated instrument method is currently the only feasible method to routinely provide rapid (3- to 6-hr) results. All of the instruments to be discussed in this chapter are capable of generating test results following such abbreviated incubation periods, although testing of some species may require longer by certain of the methods. All of the instruments function by applying a similar principle of photometric detection of bacterial growth in a broth medium. In all but one instance (detection of hydrolysis of a fluorogenic substrate) the instruments measure subtle increases in turbidity of a liquid as the bacterial mass increases during cellular multiplication. The use of sensitive optical detection systems allows recognition of slight changes in turbidity often hours before such changes could be accurately noted by the unaided eye using a manual test method.

The suppression of turbidity as evidence of the inhibitory effect of an antimicrobial agent, or conversely the increase in turbidity in the presence of a drug as an indication of microbial resistance, has been firmly established in conventional broth dilution methods (including microdilution procedures described in Chapter 6). Thus, rapid automated susceptibility test instruments merely compress the incubation/analysis period of otherwise conventional methods, with the important addition of instrument assistance in the interpretation of test results and assembly of a printed report.

The virtue in performing antimicrobial susceptibility tests in 3 to 6 hr rather than following 18 hr of conventional incubation has been frequently debated.[1,2,3] It is clear that optimal utilization of rapid results requires additional diligence on the part of the laboratory to make physicians aware of the earlier results.[4,5] Despite laboratories' need to rearrange their work flow to accomodate the same-day incubation period, and the need for physicians to slightly modify their practice habits, the emphasis on rapid testing continues to expand. This is clearly manifest in the increasing popularity of rapid, automated instruments for susceptibility testing.[6]

II. GENERAL PRINCIPLES OF RAPID SUSCEPTIBILITY TEST INSTRUMENTS

The first automated microbiology instrument to provide results after a short incubation period was ironically never marketed. The TASS system (Technicon Instruments Corp., Tarrytown, N.Y.) was a highly automated device which provided qualitative antimicrobial susceptibility results following only 3 hr of incubation.[7] The TASS system was in many ways ahead of its time, and was felt by its developers to be too complex and expensive to be a commercial success.

Most of the currently available instruments (which will be discussed separately below) rely on the same basic optical principle of paired light source-detector systems detecting increases in turbidity of bacterial suspensions (Table 1). The Autobac® system detects light scattering (detection of forward light scatter at 90° angle from the light source) due to bacterial cells, while the Abbott MS®-2/Avantage, Vitek® and COBAS-Bact (not yet marketed) systems measure turbidity as optical density (detector at 180° angle to light source). The Gibco Sensititre® Autoreader, which will be marketed shortly, is unique in its measurement of fluorescence emission from bacterial metabolism of a fluorogenic substrate during growth.

Table 1
DESCRIPTION OF HARDWARE UTILIZED IN CURRENTLY MARKETED RAPID SUSCEPTIBILITY SYSTEMS

Instrument	Computer module	Incubator-reader	No. of simultaneous tests	Principle of test analysis of susceptibility
Autobac® II (General Diagnostics, Morris Plains, N.J.)	Photometer module contains microprocessor; RS-232 interface port included; separate keyboard and IBM® PC computer, 20 megabyte Winchester hard disk, Okidata printer, and color CRT monitor	30- or 60-Place external shaker-incubators available; manual test insertion in photometer module	Because of manual cuvette insertion for interpretation, capacity is effectively unlimited	Analysis of susceptibility; comparison of final light scattering indices (LSI) for each drug with that of uninhibited control well
Abbott MS®-2/Avantage (Abbott Diagnostics Division, Irving, Tex.)	Hewlett-Packard® 9816 microprocessor with 512 K RAM (Pascal programming language), HP® 9121 D disk memory with 573 K storage on dual disk drives with 3 1/2 in. micro floppy disks, 9 in. CRT with keyboard, RS 232 port; Epson printer produces single or multipart paper reports	Analysis modules contains 8 cuvette positions for 8 simultaneous susceptibilities; up to 8 analysis modules can be coupled with one computer module	Depending upon number of analysis modules, up to 63 susceptibility tests every 4—5 hr	Kinetic growth curve analysis derived from optical density measurements taken every 5 min of drug and control wells; internal algorithms determine interpretations
Vitek® system (Vitek Systems, Inc., Hazelwood, Mo.)	Motorola® 68010 16/32 bit microprocessor, 10 megabyte Winchester hard disk, CRT and keyboard, and a multicopy printer	Reader-incubator modules have capacity for 30, 60, 120, or 240 cards in a carousel-like fashion	Depending upon size of reader-incubator module, from 30—240 susceptibility tests can be performed every 4—8 hr	Kinetic growth curve analysis derived from optical density measurements taken every 20 min of drug and control wells; internal algorithms determine interpretations

The instruments vary in their methods of exposing the bacterial test culture to the optical detector system. In the case of the Autobac®, the light source-detector passes sequentially along the sides of a 19-position plastic cuvette which has been manually inserted into the reader module of the instrument. The Vitek® system utilizes simple robotics to move thin plastic cards containing 30 small broth chambers into position for scanning by a single, fixed optical detector system. The COBAS-Bact also utilizes robotics to move round plastic rotors from the incubation to reading compartments of the instrument where the rotor is examined in stepwise increments during rotation in front of the fixed optical detection system. The Abbott MS®-2/Avantage takes the most expensive and complex approach of providing separate, matched photodetectors

Table 2
FEATURES OF CONSUMABLE SUPPLIES USED IN CURRENTLY MARKETED RAPID SUSCEPTIBILITY TEST SYSTEMS

Instrument	Test cuvette or card	Method of supplying antimicrobials for testing	Susceptibility test medium	Inoculum requirements	Automated test steps
Autobac® II	13- and 19-Position plastic cuvettes to contain 12—18 different antimicrobials or concentration increments	Antibiotics contained on filter paper disks; user selects from among different antimicrobial disks currently available	Eugon broth	Instrument-standardized inoculum prepared in Eugon broth test medium	Only the final test interpretation and printing of results (cuvettes must be incubated externally, then placed in photometer for interpretation)
Abbott MS®-2/Avantage	11-Position plastic cuvettes to contain 10 different antimicrobials or concentration increments	Antimicrobials contained on filter paper disks; user selects test batteries from among different disks currently available	IsoSensitest broth	200—400 $\mu\ell$ (Depending on whether Gram positive or Gram negative bacterium) of 0.5 McFarland standard suspension in water or saline	Incubation and final test interpretation and printing
Vitek® System	30-Position plastic cards which contain up to 11 different antimicrobials or concentration increments	Antimicrobials lyophilized in individual wells in cards; some choice of drug configuration/combinations available from manufacturer	Mueller-Hinton broth	Filler-sealer module draws 0.5—1 McFarland standard suspension in cards by vacuum	Incubation and final test interpretation and printing

for each of 12 growth chambers within a plastic cuvette which contains only one test organism. Depending upon the number of anticipated simultaneous tests, separate analysis modules must be purchased for each set of eight test sets to be performed in the Abbott instrument. The newer Avantage version may be configured with from 1 to 8 analysis modules, thus allowing from 8 to 63 simultaneous susceptibility tests (see Table 1).

Each of the three currently marketed instruments (Autobac® II, Abbott MS®-2/Avantage, and Vitek®) utilizes disposable plastic cuvettes or cards as analysis chambers to contain the organisms and antimicrobial agents during testing (see below and Table 2). In the case of the Autobac® and MS®-2/Avantage, the cuvettes are substantial moldings of 11 to 19 chambers, each chamber holding approximately 1 mℓ of growth medium plus an antimicrobial impregnated paper disk. Special elution disks are prepared by the manufacturers for configuration into test batteries of 10 to 18 drugs by individual laboratory users.

The use of paper elution disks provides somewhat more instrument flexibility than the thin plastic cards used by Vitek® to contain lyophilized antimicrobials in miniature wells. In order to provide flexibility for different medical practice environments, Vitek® produces cards with different drugs for testing of Gram-positive bacteria (called the GPS card), Gram-negative bacteria (designated the GNS card), or urinary isolates (GNU card), as well as offering to prepare custom configured cards if a laboratory will order a minimum of 5000 of the special cards.

III. SPECIFIC FEATURES OF CURRENT INSTRUMENTS

A. Autobac® Series II

The Autobac system was introduced in the early 1970s by its original manufacturer, Pfizer Diagnostics. Now made by General Diagnostics and called the Autobac® II, it holds the distinction of being the first "automated" microbiology instrument to be sold in the U.S. (since TASS was never marketed). The Autobac® was originally designed to provide rapid, qualitative susceptibility results on the most frequently tested bacteria, i.e., Enterobacteriaceae, *Pseudomonas aeruginosa,* staphylococci, and enterococci. The instrument has continued in its development to include performance of quantiative susceptibility tests, screening of urines for bacteria, and rapid identification of aerobic Gram-negative bacilli.[8] The latter function is accomplished using the innovative approach of growth inhibition profiles with a group of unique dyes and chemicals as well as a few conventional antimicrobial agents.[9] The MIC capability of the instrument is based on linear regression analysis of light scatter indices obtained from testing a high and low concentration of each drug. These computer-calculated MICs have been shown to compare favorably with traditionally derived values.[10]

The more popular qualitative mode (S, I, or R categories) of the Autobac® has been extensively evaluated in a number of studies. In fact, a large multicenter (seven-laboratory) collaborative study[11] set the trend for all such evaluations of future instruments or methods. The rationale was that by simultaneous evaluation of the instrument using a common protocol by laboratories in a number of different geographic areas, an indication of the performance of an instrument in actual day-to-day situations is likely to result. This landmark publication also provided new jargon for evaluation of subsequent susceptibility test data, i.e., the terms "minor," "major," and "very major" errors. These terms were coined to describe discrepancies between an automated test method and a reference susceptibility method such as agar dilution or the Kirby-Bauer disk diffusion method. A discrepancy was considered minor if an intermediate category result was obtained by one method and either a susceptible or resistant result by the other method. A major discrepancy was declared when an organism was said to be resistant by the automated system but susceptible by the reference procedure. A very major error occurred when the instrument result indicated susceptibility and the reference method defined resistance of a microorganism.

Use of these terms has since become commonplace to describe the types of discordance between newer automated and older reference susceptibility test procedures. However, consensus guidelines for determining acceptable levels of the various types of errors has not been reached. Thornsberry[12] has suggested that complete correlation of a new method with an established procedure should be over 90%, and the combined major and very major errors of the new instrument should be less than 5%. Others[13] have suggested more stringent criteria, including the recommendation that total errors should not exceed 5% and that very major errors should not be greater than 1.5%.

In the initial collaborative evaluation of the Autobac® there was greater than 90% agreement between the instrument-generated results and those obtained by agar dilution and disk diffusion testing.[11] However, a number of test errors occurred with certain drug-organism combinations, e.g., *Enterobacter* spp. with nitrofurantoin and tetracycline; *Serratia* spp. with polymyxin and colistin; *Pseudomonas aeruginosa* with chloramphenicol, kanamycin, and gentamicin; enterococci with tetracycline, kanamycin, and gentamicin; and *Staphylococcus epidermidis* with penicillin and tetracycline. One of these discrepancies was subsequently remedied, i.e., *Proteus* spp. testing with nitrofurantoin. However, some errors were not remedied and have been contraindicated for testing by the manufacturer, i.e., enterococci with cephalothin and *Pseudomonas aeruginosa* with kanamycin. Later studies also defined problem combinations including *S. aureus* with erythromycin[14] and methicillin,[15,16] and *Escherichia coli* with ampicillin.[17] A recent study[18] documents that the Autobac® continues to be unreliable for detection of methicillin resistant *S. aureus*. This latter problem is probably due to the 37°C incubation temperature of the Autobac® which has been shown to be unfavorable for detection of methicillin heteroresistance in several studies.[19,20] However, if these manufacturer-disclaimed or unfavorable organism-drug combinations are excluded, the Autobac® provides accuracy equivalent to or superior to other rapid methods, i.e., ≥90% compared with conventional methods.[11,12,21]

B. Abbott MS®-2/Avantage

The Abbott MS®-2 system was introduced a few years after the Autobac®, and in many ways seems to represent an improved and more automated execution of a similar concept. Instead of using a light scatter index, the MS®-2 determines the more conventional optical density at a fixed wavelength (670 nm) suitable for measurement of bacterial cells in an amber-colored growth medium. An increase in automation occurred when individual light emitting diodes (LEDs) and detectors were provided for each test position. After insertion of a cuvette in the instrument, and identifying the test to be performed by use of the keyboard, all subsequent operations are performed automatically. In addition, the MS®-2 makes an internal inoculum adjustment by growing the bacteria for a brief period to establish logarithmic growth prior to exposure of the culture to the test antibiotics. This is accomplished by use of a two-tiered plastic cuvette (the upper chamber being the inoculum growth area) and a series of pumps to pressurize the upper chamber and cause the inoculated broth to transfer to the bottom antibiotic-containing chambers. The optical density readings in the antibiotic disk chambers are then determined every 5 min in the ensuing 3 to 6 hr (depending on the organism's growth rate).

The MS®-2 prints a qualitative S, I, or R result at the conclusion of the test, and in addition, prints an MIC breakpoint for S or R results, e.g., susceptible, ≤10 μg/mℓ, or resistant, ≥30 μg/mℓ. The range for the S and R interpretations extends from the antibiotic content of the elution disk (for susceptible) to three times the disk concentration (for resistant). For intermediate results, a programmatically calculated MIC value is provided.

A popular option for the MS®-2 (dubbed the Research system) was the addition of a CRT, hard-copy printer, and the software necessary to display plots of the growth curves during the tests' incubation. This feature has been incorporated into the successor of the MS®-2, called the Avantage microbiology center.[22] The Avantage incorporates all of the same principles of the MS®-2 (including use of the same analysis modules and consumable supplies, Tables 1 and 2). However, the MS®-2 control module was replaced by a Hewlett-Packard® 9816 computer with expanded capabilities. The additional computer memory allows storage of more information from simultaneous tests on the same bacterial isolate (identification and susceptibility) and programmatically compiles these results into a final report. A CRT and graphics capability also has become a standard feature of the Avantage.

The MS®-2/Avantage, like the Autobac®, has been carefully scrutinized for reproducibility and accuracy in a number of published studies. The MS®-2 system was evaluated by a four-center collaborative study[23] which again introduced more comprehensive methods for testing of new instruments. In addition to testing a large number of clinical bacterial isolates in each laboratory, a set of "challenge" strains selected for their unusual resistance patterns was distributed to each laboratory to assure a thorough exploration of the instrument's performance. These challenge strains included *S. aureus* resistant to methicillin and strains demonstrating inducible erythromycin resistance, both of which were unusual in the U.S. at the time the study was performed.

As with the Autobac®, several organism-drug combinations were recognized as problematic for the MS®-2, e.g., *Enterobacter* spp. with ampicillin and cephalothin, *Serratia* spp. with colistin, and enterococci with penicillin, cephalothin, kanamycin, and gentamicin.[23] Subsequent studies have confirmed and extended these original observations regarding the MS®-2.[24-27] One group more recently reported numerous very major errors when testing two newer cephalosporins (cefotaxime and moxalactam) against *P. aeruginosa*.[28]

Despite the initial challenge of the MS®-2 with methicillin-resistant strains of *S. aureus* from Europe,[23] several laboratories reported difficulties in recognition of the heteroresistant strains which became prominent in some U.S. hospitals in the mid-1970s.[16,29,30] These reports led Abbott to revise the computer software used to interpret methicillin tests with a subsequent improvement in accuracy when resistant strains from several different areas of the country were tested.[31] When the few unfavorable organism-drug combinations described above are excluded, the accuracy of the MS®-2/Avantage (compared to an 18- to 24-hr reference test method) is quite similar to that of the Autobac®, i.e., greater than 90% accuracy.[21,23,27]

C. The Vitek® System

The Vitek® system (formerly called the AutoMicrobic® system or AMS®) is a by-product of the aerospace industry. It was originally designed by the McDonnell-Douglas Corp. for the National Aeronautics and Space Administration (NASA) as an on-board test system for spacecraft exploring other planets for life. Because of its intended use by nonmicrobiologists, it was highly automated from its inception. Very small plastic cards were originally configured to contain biochemical tests and selective growth media for organism detection and identification.

The Vitek® system continues to be marketed by a subsidiary of McDonnell-Douglas, Vitek® Systems Inc. For determination of antimicrobial susceptibility, the Vitek® originally performed a qualitative test with S, I, or R results on only *Escherichia coli* or *Proteus mirabilis*. More recently, "MIC" cards have been developed (GNS, GNU, or GPS cards) which allow quantitative results on most rapidly growing aerobic bacteria in 3 to 10 hr.

Like the MS®-2/Avantage, the Vitek® uses kinetic measurements of growth in the presence of antibiotics (every 20 min with the Vitek®) to provide linear regression analyses and ultimately calculated MIC values. The Vitek®, in fact, offers the operator a choice of interpretive qualitative or quantitative formats for results. Despite the fact that the Vitek® was approved by the U.S. FDA and has become perhaps the most popular of the instruments described in this chapter, its susceptibility test function has never been subjected to the close scrutiny of a large multicenter collaborative study as performed with the Autobac® and MS®-2 systems. When this is coupled with the fact that Vitek® has provided numerous modifications to both the test cards and the computer software used for their interpretation during the period that the instrument has been marketed, it is difficult to ascertain the current level of accuracy of the instrument's results. Independent studies have demonstrated a lower level of accuracy of the Vitek® qualitative test results than those obtained by the Autobac® or MS®-2,[21,26] or poor correlation of the Vitek® MIC results (not always within ±1 twofold concentration agreement of a reference microdilution procedure).[27,32,33] Unfavorable organism-drug combinations include testing of *Enterobacter* spp., *Serratia* spp., and *Providencia* spp. with ampicillin, cefamandole, and nitrofurantoin;[34-36] *Citrobacter* spp. with ampicillin;[37] *Proteus mirabilis, Klebsiella* spp., *Enterobacter* spp., and *Citrobacter* spp. with chloramphenicol;[35,37] and most importantly, *P. aeruginosa* with gentamicin, amikacin, and carbenicillin.[27,35,37] However, the most recently published studies seem to suggest that the accuracy of test results has been improved in recent years.[36-38] In addition, the Vitek® has been shown by several workers to be the most accurate rapid test system with respect to detection of methicillin-resistant *S. aureus*.[31,39] However, few data exist on the accuracy of the GPS card with other species and antimicrobials.

As mentioned previously, the fixed format of the Vitek® cards provides some limitations in the selection of drugs for testing. However, the large number of wells in the cards combined with the most commonly used drugs being present in the several standard cards, provides an acceptable selection for most users. Large laboratories with special needs may choose to have their cards custom configured by Vitek®.

Due in large part to its aerospace heritage and the simple fact that individual cards are identified as to accession numbers and task to be performed in machine read markings, the Vitek® is perceived by many to be the most automated of the three current systems. Supported by a large and active Vitek® users group, the instrument is the choice of many knowledgable microbiologists. Previous criticisms of lower accuracy than that of either the Autobac® or MS®-2 may no longer apply in light of continued updates and improvements by the manufacturer. Perhaps the main continuing performance advantage of the Autobac® or MS®-2/Avantage is that of greater speed, since Vitek® results may require 6 to 10 hr total incubation with as many as 40% of bacterial strains. Many microbiologists undoubtedly choose the Vitek® for its labor-saving level of automation rather than as an instrument to produce rapid results.

IV. INSTRUMENTS CURRENTLY UNDER DEVELOPMENT

The introduction of newer automated instruments for susceptibility testing has not occurred in nearly a decade. The adaptation of the Vitek® system for performance of susceptibility tests represented, until recently, the last instrument to be marketed in the U.S. for rapid susceptibility testing. Gibco Laboratories has recently revealed a modified microdilution system which allows final test readings to be taken after 5 hr of incubation through use of an automated tray reader (Sensititre® Autoreader) which detects release of a fluorescent substrate label.[40] If a bacterium is allowed to grow at the expense of the labeled substrate (and thus is not inhibited by particular concentrations of an antibiotic), the instrument detects the emission of fluorescence as an indication of bacterial growth. This technique allows a number of antimicrobials in different concentration increments to be tested very conveniently (as is the case with microdilution tests which require overnight incubation; see Chapter 6). In addition, the user can choose prior to inoculation whether to employ the rapid 5-hr readings or to incubate the trays for a conventional 18-hr period before instrument interpretation. While Gibco also provides an automated device for tray inoculation, several manual steps are required for incubation and final reading of the trays. Since this instrument has only recently been unveiled, relatively few data are available to assess its accuracy and technical facility.[40,41] However, it appears that another manufacturer of conventional microdilution test systems, American Micro Scan, may employ a very similar approach to rapid interpretation of microdilution tests by use of a fluorescent substrate.

The COBAS-Bact is a highly automated microbiology instrument designed by Roche Diagnostics in Europe. As described previously, the Bact is a sophisticated centrifugal analyzer which at first glance would seem to be superior to currently marketed instruments. However, limited data derived in this country have suggested considerably lower accuracy than that described above with currently FDA-approved and marketed instruments.[42] Moreover, the fixed configuration of the Bact allows only 50 simultaneous tests (including identification and susceptibility tests) which would not be adequate for many of the large laboratories which constitute the most likely purchasers of such an instrument. It does not appear that the instrument's capacity can be increased without substantial, expensive redesigning. At this writing it seems unlikely that the COBAS-Bact will be marketed in the U.S. in its current form.

It is not clear whether other new instruments are nearing the final phases of development at this time. The manufacturers of the three instruments which have been described in greatest depth in this chapter do not appear ready with more modern replacements for their aging designs. Thus, it seems that only slight refinements can be expected for the Autobac®, Avantage, and Vitek® in the foreseeable future. Instead, microdilution systems which have heretofore required overnight test incubation may be successfully modified to allow rapid 4- to 5-hr readings through the use of more sensitive means of growth detection, as in the case of the Gibco Sensititre® Autoreader. However, since these will be designed to utilize existing conventional microdilution trays, they will likely be large, mechanically complex instruments. An ultimately more favorable approach would seem to include design of small sample holders which can be easily handled using robotics and which do not require the addition of reagents following incubation, much as the current Vitek® instrument has successfully employed.

V. ADVANTAGES OF CURRENT SYSTEMS

The instrument systems described in this chapter probably should be viewed as the adolescent period of automation in microbiology. All of these systems offer the potential for improved reproducibility of susceptibility tests and provide more rapid results than can conveniently be achieved by any manual procedure. Most evaluations have reported that the instruments function reliably in a mechanical sense and are not difficult for the majority of medical technologists to operate. Certain of these systems appear to offer labor savings over manually performed tests. All of the instruments can accomplish the most commonly performed functions in the microbiology laboratory, i.e., organism identification and antimicrobial susceptibility testing. Therefore, the potential currently exists to automate a sizable portion of a laboratory's daily work load using one or more of these instruments. In addition, data management capability can be added to the instruments' operating computers with any of the current systems.[22]

VI. DISADVANTAGES OF CURRENTLY AVAILABLE SYSTEMS

The instruments discussed in this chapter (including those still under development) offer automation of several of the steps involved in performing an antimicrobial susceptibility test, but by no means is the entire procedure performed by any of the instruments. The microbiologist must still isolate bacteria using conventional culture methods and prepare a bacterial suspension of defined density for testing by the instrument. Thus the "automated" aspects of the test do not begin until after the manual inoculum preparation. Indeed, inadequacy of inoculum preparation is a major variable affecting performance of all of the instruments. Moreover, none of the existing instruments attempt to analyze direct patient samples for the possible presence of resistant bacteria.

Because of instrument control of a number of steps in the test procedure, instrument methods may be expected to offer some improvement in test reproducibility.[43] However, questions of rapid test accuracy continue to be debated.[12,14,21,27,29,30,37] The very nature of the short incubation period employed by the rapid test instruments causes some compromise in accuracy compared to overnight methods.[12] An incubation period of only 3 to 6 hr may not always be adequate for expression of some bacterial resistance mechanisms, e.g., inducible β-lactamases,[44] high mutation rates to drug resistance,[14] or heterogenous population characteristics.[18] Thus, overnight incubation of susceptibility tests may provide a better opportunity to detect β-lactamase mediated resistance in staphylococci or some Gram-negative bacilli,[14,44] and erythromycin resistance or methicillin heteroresistance in staphylococci.[14,16] Some of these discrepancies have been resolved in large part by manipulating the inoculum density (higher for the rapid test systems), by altering the drug test concentrations, by modifying the computer software for interpretation of results, or by adding inducer drugs to the test medium to promote resistance expression in a shorter time period.[12,44] Despite these improvements, nagging disagreements still occur between rapid automated and overnight test results. The criteria for acceptability of results described earlier is based on the observation that error rates of 5 to 10% are unavoidable by current methodology. Thus, a conscientious microbiologist must be knowledgable of unfavorable organism-drug combinations and provide alternative accurate means for their testing.

Additional problems include relevant performance of quality control and provision for instrument backup in case of mechanical failures. All current instruments for susceptibility testing are mechanically and optically complex devices which must function properly for reproducible test results. Each manufacturer describes routine maintenance and function checks to prevent or detect overt malfunctions. However, it has been aptly pointed out that the quality control measures suggested by the manufacturers are inadequate to detect subtle deterioration in potencies of the test antibiotics since the control microorganisms used to verify test function are usually either highly susceptible or resistant to the test antimicrobials.[45,46] Lack of control precision due to these off-scale values has not yet been satisfactorily addressed by the manufacturers.

When a mechanical failure does occur, most of the instrument methods cannot be performed using manual interpretive methods since subtle and varying definitions of growth are employed by the instruments. Only the Gibco Sensititre® microdilution-based method offers the feasibility of manual test interpretation after an extended incubation period, i.e., 18 to 24 hr. Therefore, laboratories must always maintain some different backup testing procedure such as Kirby-Bauer or overnight microdilution testing for the Autobac®, MS®-2/Avantage and Vitek® instrument systems.

Lastly, nonmicrobiologists may be surprised to learn that automated susceptibility test instruments generally do not provide direct cost savings over manually performed tests. Unlike the greater economy afforded by automated test procedures performed by chemistry autoanalyzers, consumable costs for microbiology instruments are equivalent or greater than those for manual tests. In addition, the instruments described in this chapter cost from $30,000 to $90,000 for hardware depending upon the brand of instrument and number of simultaneous functions to be performed. Thus, many laboratories choose to obtain an instrument system using a reagent rental or lease agreement to avoid the large capital expenditure for sophisticated hardware which is constantly evolving. It is unlikely that cost-effectiveness of these instruments can be demonstrated unless labor savings and ultimate medical benefits derived from rapid reporting can be documented.[47] While an improvement in patient care has been suggested when rapid results are employed,[1,48] accurate quantitation of such benefits is difficult to obtain.

VII. SUMMARY

Instruments for performance of rapid antimicrobial susceptibility testing should be viewed as one of the best examples of automation in microbiology. The performance of susceptibility tests seems to be well suited to instrumentation, since a finite determination of microbial properties is being sought in the test. Instruments provide more rapid results than can be generated using manual systems and undoubtedly serve to improve intra- and interlaboratory reproducibility and standardization.

Loyal users of one or another instrument system may often debate which of the available machines is the most "automated". It seems clear that the Autobac® system requires the greater number of manual operations, while both the MS®-2/Avantage and Vitek® systems provide more machine-accomplished functions. Indeed, the Vitek® system, with its aerospace ancestry may offer the highest level of automation currently available, especially if its other functions, i.e., bacterial and yeast identification, and urine screening are considered. The newer instruments undergoing their latter phases of development (Sensititre® Autoreader, COBAS-Bact, and American Micro Scan Walk-Away) do not appear likely to offer more automated features than those described above. The advanced automated features of Technicon's ill-fated TASS have still gone unmatched by its more "modern" successors. The instruments herein represent only the beginning of automation in microbiology. They portend of more sophisticated and labor-saving instruments yet to be developed. Undoubtedly, further advances in optical, electronic, and computerized data handling methods will yield even more attractive instruments in the near future.

REFERENCES

1. Matsen, J. M., Rapid reporting of results — impact on patient, physician, and laboratory, in *Rapid Methods and Automation in Microbiology,* Tilton, R. C., Ed., American Society for Microbiology, Washington, D.C., 1982, 98.
2. Ellner, P. D., Rapid reporting — does it really affect patient care?, in *Rapid Methods and Automation in Microbiology,* Tilton, R. C., Ed., American Society for Microbiology, Washington, D.C., 1982, 106.
3. Doern, G. V., Clinical impact of rapid susceptibility testing, *Diagn. Microbiol. Infect. Dis.,* 3, 59S, 1985.
4. Matsen, J. M., Means to facilitate physician acceptance and use of rapid test results, *Diagn. Microbiol. Infect. Dis.,* 2, 73S, 1985.
5. Ryan, K. J., Methods for rapid transmission of microbiology reports, *Diagn. Microbiol. Infect. Dis.,* 2, 33S, 1985.
6. Jones, R. N., Antimicrobial susceptibility testing (AST): a review of changing trends, quality control guidelines, test accuracy, and recommendation for the testing of β-lactam drugs, *Diagn. Microbiol. Infect. Dis.,* 1, 1, 1983.
7. Isenberg, H. D., Reicher, A., and Wiseman, D., Prototype of a fully automated device for determination of bacterial antibiotic susceptibility in the clinical laboratory, *Appl. Microbiol.,* 22, 878, 1971.
8. Malloy, P. J., Miceika, B. G., and Ducate, M. J., Automated methods in microbiology. II. Identification and susceptibility, *Am. J. Med. Technol.,* 49, 313, 1983.
9. Sielaff, B. H., Matsen, J. M., and McKie, J. E., Novel approach to bacterial identification that uses the Autobac system, *J. Clin. Microbiol.,* 15, 1103, 1982.
10. Schoenknecht, F. D., Washington, J. A., II, Gavan, T. L., Thornsberry, C., Rapid determination of minimum inhibitory concentrations of antimicrobial agents by the Autobac method: a collaborative study, *Antimicrob. Agents Chemother.,* 17, 824, 1980.

11. Thornsberry, C., Gavan, T. L., Sherris, J. C., Balows, A., Matsen, J. M., Sabath, L. D., Schoenknecht, F., Thrupp, L. D., Washington, J. A., II, Laboratory evaluation of a rapid, automated susceptibility testing system: report of a collaborative study, *Antimicrob. Agents Chemother.*, 7, 466, 1975.
12. Thornsberry, C., Automated procedures for antimicrobial susceptibility tests, in *Manual of Clinical Microbiology*, 4th ed., Lennette, E. H., Balows, A., Hausler, W. J., Jr., and Shadomy, H. J., Eds., American Society for Microbiology, Washington, D.C., 1985, chap. 107.
13. Sherris, J. C. and Ryan, K. J., Evaluation of automated and rapid methods, in *Rapid Methods and Automation in Microbiology*, Tilton, R. C., Ed., American Society for Microbiology, Washington, D.C., 1982, 1.
14. Lampe, M. F., Aitken, C. L., Dennis, P. G., Forsythe, P. S., Patrick, K. E., Schoenknecht, F. D., and Sherris, J. C., Relationship of early readings of minimal inhibitory concentrations to the results of overnight tests, *Antimicrob. Agents Chemother.*, 8, 429, 1975.
15. Cleary, T. J. and Maurer, D., Methicillin-resistant *Staphylococcus aureus* susceptibility testing by an automated system, Autobac I, *Antimicrob. Agents Chemother.*, 13, 837, 1978.
16. Aldridge, K. E., Janney, A., Sanders, C. V., and Marier, R. L., Interlaboratory variation of antibiograms of methicillin-resistant and methicillin-susceptible *Staphylococcus aureus* strains with conventional and commercial testing systems, *J. Clin. Microbiol.*, 18, 1226, 1983.
17. Stubbs, K. G. and Wicher, K., Laboratory evaluation of an automated antimicrobial susceptibility system, *Am. J. Clin. Pathol.*, 68, 769, 1977.
18. de la Roy, Y. R., Chevalier-Burbaud, E., Pannetier, S., and Souchaud, F., Use of the Autobac system for detection of methicillin-resistant *Staphylococcus aureus*, *J. Clin. Microbiol.*, 22, 467, 1985.
19. Barber, M., Naturally occurring methicillin-resistant staphylococci, *J. Gen. Microbiol.*, 35, 183, 1964.
20. Thornsberry, C., Caruthers, J. Q., and Baker, C. N., Effect of temperature on the *in vitro* susceptibility of *Staphylococcus aureus* to penicillinase-resistant penicillins, *Antimicrob. Agents Chemother.*, 4, 263, 1973.
21. Hansen, S. L. and Freedy, P. K., Concurrent comparability of automated systems and commercially prepared microdilution trays for susceptibility testing, *J. Clin. Microbiol.*, 17, 878, 1983.
22. Blatt, M. L., Automated microbiology systems, *J. Clin. Lab. Automat.*, 4, 64, 1984.
23. Thornsberry, C., Anhalt, J. P., Washington, J. A., II, McCarthy, L. R., Schoenknecht, F. D., Sherris, J. C., and Spencer, H. J., Clinical laboratory evaluation of the Abbott MS-2 automated antimicrobial susceptibility testing system: report of a collaborative study, *J. Clin. Microbiol.*, 12, 375, 1980.
24. Barnes, W. G., Green, L. R., and Talley, R. L., Clinical evaluation of automated antibiotic susceptibility testing with the MS-2 system, *J. Clin. Microbiol.*, 12, 527, 1980.
25. Sewell, D. L. and Makler, M. T., Clinical evaluation of the Abbott MS-2 antimicrobial susceptibility testing system, *Am. J. Clin. Pathol.*, 76, 82, 1981.
26. Kelly, M. T., Latimer, J. M., and Balfour, L. C., Comparison of three automated systems for antimicrobial susceptibility testing of gram-negative bacilli, *J. Clin. Microbiol.*, 15, 902, 1982.
27. Johnson, J. E., Jorgensen, J. H., Crawford, S. A., Redding, J. S., and Pruneda, R. C., Comparison of two automated instrument systems for rapid susceptibility testing of gram-negative bacilli, *J. Clin. Microbiol.*, 18, 1301, 1983.
28. Stone, L. L. and Jungkind, D. L., False-susceptible results from the MS-2 system used for testing resistant *Pseudomonas aeruginosa* against two third-generation cephalosporins, moxalactam and cefotaxime, *J. Clin. Microbiol.*, 18, 389, 1983.
29. Boyce, J. M., White, R. L., Bonner, M. C., and Lockwood, W. R., Reliability of the MS-2 system in detecting methicillin-resistant *Staphylococcus aureus*, *J. Clin. Microbiol.*, 15, 220, 1982.
30. Carlson, J. R., Conley, F. E., and Cahall, D. L., Methicillin-resistant *Staphylococcus aureus* susceptibility testing with the Abbott MS-2 system, *Antimicrob. Agents Chemother.*, 21, 676, 1982.
31. Jorgensen, J. H., Redding, J., Johnson, J. E., Holloway, V., and Almeida, R. J., Rapid recognition of methicillin-resistant *Staphylococcus aureus* by use of automated test systems, *J. Clin. Microbiol.*, 20, 430, 1984.
32. Dudley, M., Nauscheutz, F., and Juchau, S. V., Clinical evaluation of new AMS GSC Plus card, *Diagn. Microbiol. Infect. Dis.*, 1, 139, 1983.
33. Woolfrey, B. F., Lally, R. T., Ederer, M. N., and Quall, C. O., Evaluation of the AutoMicrobic system for susceptibility testing of *Pseudomonas aeruginosa* to gentamicin, tobramycin, and amikacin, *J. Clin. Microbiol.*, 19, 502, 1984.
34. Goldstein, J., Guarneri, J. J., and Scherer, J., Use of the AutoMicrobic system for rapid antimicrobial susceptibility testing of Enterobacteriaceae in a clinical laboratory, *J. Clin. Lab. Automat.*, 2, 329, 1982.

35. Backes, B. A., Cavalieri, S. J., Rudrick, J. T., and Britt, E. M., Rapid antimicrobial susceptibility testing of gram-negative clinical isolates with the AutoMicrobic system, *J. Clin. Microbiol.*, 19, 744, 1984.
36. Keiber, N. E., Kelly, M. T., Latimer, J. M., Tison, D. L., and Hysmith, R. M., Comparison of the Cathra Repliscan II, the AutoMicrobic System Gram-Negative General Susceptibility-Plus card, and the Micro-Media System Fox panel for dilution susceptibility testing of gram-negative bacilli, *J. Clin. Microbiol.*, 21, 959, 1985.
37. Nadler, H. L., Dolan, C., Mele, L., and Kurtz, S. R., Accuracy and reproducibility of the AutoMicrobic System Gram-Negative General Susceptibility-Plus card for testing selected challenge organisms, *J. Clin. Microbiol.*, 22, 355, 1985.
38. Woolfrey, B. F., Lally, R. T., Ederer, M. N., and Quall, C. O., Evaluation of the AutoMicrobic System for identification and susceptibility testing of gram-negative bacilli, *J. Clin. Microbiol.*, 20, 1053, 1984.
39. Stotler, R. W. and Meyer, M. C., Detection of oxacillin resistant Staphylococci by the AutoMicrobic System, *J. Cin. Microbiol.*, 18, 1205, 1983.
40. Staneck, J. L., Allen, S. D., Harris, E. E., and Tilton, R. C., Automated reading of MIC microdilution trays containing fluorogenic enzyme substrates with the Sensititre Autoreader, *J. Clin. Microbiol.*, 22, 187, 1985.
41. Staneck, J. L., Allen, S. D., and Tilton, R. C., Rapid MIC determinations using the Sensititre Autoreader, *Abstr. Annu. Meet. Am. Soc. Microbiol.*, American Society for Microbiology, Washington, D.C., 1985, Abstr. C89.
42. Murray, P. R., Hom, R., Martin, W., and Rupka, G., *In vitro* collaborative evaluation of the COBAS-Bact antimicrobial susceptibility instrument, *Abstr. Annu. Meet. Am. Soc. Microbiol.*, American Society for Microbiology, Washington, D.C., 1985, Abstr. C82.
43. Murray, P. R., Antibiotic susceptibility testing, Part II, *Lab. Med.*, 14, 417, 1983.
44. Jorgensen, J. H.,. Susceptibility test methods which may be able to predict the emergence of resistance to newer β-lactam antibiotics, *Chemioterapia*, 4, 7, 1985.
45. Boshard, R. and Matsen, J. M., Alternative quality control parameters for Autobac susceptibility testing disks: use of agar diffusion zone size results, *J. Clin. Microbiol.*, 13, 814, 1981.
46. Kellogg, J. A., Inability to adequately control antimicrobial agents on AutoMicrobic System gram-positive and gram-negative cards, *J. Clin. Microbiol.*, 21, 454, 1985.
47. Ferraro, M. J., Potential impact of rapid microbiology tests in the prospective payment era, *Diagn. Microbiol. Infect. Dis.*, 2, 65S, 1985.
48. Jorgensen, J. H., Clinical situations in which rapid methods may best be applied, *Diagn. Microbiol. Infect. Dis.*, 2, 3S, 1985.

Immunologic Methods for Diagnosis of Infectious Diseases or Measurement of Antimicrobial Agents in Body Fluids

Chapter 9

AUTOMATION OF ANTIGEN DETECTION IN INFECTIOUS DISEASE DIAGNOSIS

F. A. Meier and H. R. Hill

TABLE OF CONTENTS

I. OVERVIEW OF AUTOMATION IN ANTIGEN DETECTION[1]

The role of automation in antigen detection may be subdivided according to the phases of the antigen detection process and the reasons for using an automated procedure. The four phases of the antigen detection process are preparation and dilution of the samples and reagents, separation of unbound reagents from products ("washes"), detection of antigen-antibody complexes, and data analysis of the signal generated by antigen recognition. The motives for using an automated system are increasing the speed and improving the reproducibility of the procedure. Computers play a role in each of the four phases, and the increasing availability of these machines has been a major stimulus to automation. In the first three phases, computers are operating systems that control arrays of tubes and calibrated valves for sample preparations and dilutions, tubes and calibrated manifolds for washes, and a variety of sensors for receiving the signal of the antigen recognition. The role of the computer in the fourth phase is the mathematical interpretation of the signal received by the sensing device. Computers perform the calculations that transform the received signals into quantitations of the presence of an antigen. Computer programs also arrange these data conveniently in tabular and graphic forms so that the data can be analyzed for the power of the statistical inference. Before considering the influence of automation in each of the phases of the antigen detection process and the consequences of automation in specific test formats, we will first consider the reasons for automation.

A. Reasons for Automation

If an information engineer asks the question, "What can be automated in the antigen detection process?",[2] we can respond, "In principle, everything." The more practical question is "Why automate the antigen detection process either partially or completely?" The answer appears if one examines the usual technical indices of a test's performance, chemical sensitivity, chemical specificity, reproducibility, and recovery.[3,4] One finds that the first two steps in the antigen detection process — sample preparation and dilution and reagent/product separation — dominate a test's technical reproducibility. If an automated device can achieve better reproducibility than direct human control, then automation of these steps may be justified.

An automated device often achieves better reproducibility in cases where many specimens are processed with repetitious maneuvers which must be precisely identical. Direct human control of precise repetition also tends to be slower than automated control at the same level of precision. Thus, in laboratory situations where reproducibility is important, especially where there are large numbers of specimens to process, automation may provide both greater reproducibility and faster processing.

This first justification of automation based on the number of tests performed introduces the next question, "How many specimens are necessary to justify automation on this basis?" The critical volume depends on two practical measures: the batch size (number of samples) of a run (a single complete operation of the procedure) and the turnaround time (duration from the ordering of a test to its result) required in the specific clinical situation. In practice, clinical laboratories that receive enough specimens to fill a batch during a single full day can provide a report the next day (overnight service). When specimen volume is even larger, sufficient to permit two runs per day, the laboratory can produce return of results in less than 24 hr (same day service). For almost all antigen detection tests, the volumes of samples needed to achieve these fast turnaround times are higher than those which are received by hospital laboratories. As the economics of medical laboratory testing encourages the "clustering" of laboratory services, however, centralized laboratories may be able to achieve test volumes sufficient to warrant automation of antigen testing on the basis of specimen volume alone.

One can also ask, "Is rapid turnaround a necessary requirement of automated systems?" It is, if information from the laboratory tests is being sought to establish a patient's prognosis or to establish a basis for therapy. If, however, antigen tests are being used for surveillance, screening, or confirmation of diagnoses already arrived at, then weekly, fortnightly, or even monthly batching may be medically acceptable. It is in these more leisurely time frames that automated antigen tests find themselves in reference laboratories. For these indications, automation may be justified on the basis of greater sensitivity and reproducibility without reference to test volume.

B. Sensitivity and Precision

Chemical sensitivity and reproducibility are indeed most important in screening and confirmatory tests. High technical performance is demanded in screening and confirmation to avoid economically wasteful misclassification (in the case of screening) or diagnostically confusing misclassification (in the case of confirmation). Indeed, at very low test volumes, the test operator's inexperience with a seldom performed procedure may so interfere with test reproducibility that the better performance of automated methods may override the high reagent costs of running one or two sample batches in which controls outnumber specimens.

C. Other Practical Considerations[3,4]

Besides test volume and the reproducibility of automated assays, there are three other practical considerations regarding automated methods for antigen detection. First, improved reproducibility accrues mainly from automatic operation of the process, not automated data analysis. This may change with newer sensor systems, but presently automated interpretation chiefly increases convenience and speed rather than reproducibility.

Second, although automation of an antigen test stresses different aspects of the antigen detection process than those stressed by manual methods, attention to detail is still vital to success.[5] Running an automated test takes less sophistication on the test operator's part, but "troubleshooting" problems in the automated method requires more sophistication than with a manual method. Antigen detection remains a process with many variables whether it is done manually or automatically. Consequently, quality control remains an important consideration in automated antigen detection methods despite their greater capacity for reproducibility. Better reproducibility remains a potential benefit of an automated method, not a "given" insured by the purchase of a machine.

Third, although some clinical workers have the technical background necessary to adapt components to a particular antigen detection task, most do not. Reagents, sensing devices, computers, and the programs that run them are more often sold as a package. The consumer usually depends on a single supplier for adequate immune reagents, appropriate indicator reagents, sensitive yet easily maintained sensing devices, adequate computers, and accurate control and interpretive programs. This dependence may turn the whole antigen detection operation into a "black box", similar to the kits and automated routines that dominate clinical chemistry in most hospital laboratories. Automation, thus, may transfer most of the quality control of the method from its user to its manufacturer.[3,5,6]

D. The Structure of Automated Antigen Testing

With the above practical benefits and limitations in mind, let us now consider how the components of antigen detection tests (immune reagents, indicator reagents, sensing device, and data interpreter) influence the function of automated methods.

1. Immune Reagents

Immune reagents include (1) labeled antigen (used in competitive assays), (2) antigen-specific detector antibodies, (3) anchor antibodies fixed to the solid phase in phase-separation antigen detection tests, and (4) labeled antibodies with specificity for the class, type, or species of the detector antibody used in indirect ("sandwich") techniques. The immune affinities and absorbances of these reagents are the major determinants of an antigen detection test's chemical specificity.

2. Indicator Reagents

The indicator reagents couple the antigen-antibody complex to the method's sensing device by generating the signal that the device receives. Most receivers sense some sort of radiation, so indicators either change the medium through which the radiation passes or change generation of the radiation. An example of the first sort of indicator is found among the enzyme-linked immunosorbent assays.[7] One reactant in an enzyme-substrate complex is attached either directly or indirectly to the antigen. The addition of the other reactant changes the optical density of the medium through which light passes on its way to a sensor. An example of the second sort of indicator is a fluorescent molecule attached to a detector antibody that changes some characteristic of its fluorescence (its intensity,[8] phase,[9] or polarization[10]), when the detector antibody binds specifically to the antigen.[11] The quality of the signal generated by such indicators has a major effect on the chemical sensitivity of antigen detection systems.

3. The Sensory Device[12,13]

In general, a fully automated sensor can differentiate signal strengths more precisely than direct human quantitation. This has led to the employment of automation in the sensing device in most antigen detection techniques that strive for chemical sensitivity in the clinical laboratory environment.

Since the sensing device may be a major capital expense, it often must function in a variety of antigen, antibody, or chemical analyte-detection tests. For example, one vendor of antigen detection tests supplies a sensing device designed to be used in a variety of antibody and antigen detection tests for viral antibody, and viral, chlamydial, and bacterial antigen. The same sensor can also be modified to detect biochemical reactions used to identify bacteria. Sensor versatility can be provided in three ways: (1) by making the sensor's signal-receiving range easy to modify, (2) by varying the indicator sending the signal, or (3) by changing the canons for interpreting the received signal. The availability of interchangable computer programs in cassettes makes the third modification easy and marketable.

4. Data Interpreter

The sensor's automated data acquisition can be combined with automated data interpretation by a computer program. In this context, it is, however, important to remember the computer adage "garbage in/garbage out": a discriminating sensor and elegant interpretative program will not be able to make silk-purse results from a sow's ear of inadequate immune and indicator reagents. Automated diluters and (where they are needed in phase heterogeneous techniques) automated washers often have a greater effect on how well an automated antigen test functions than do the sensor and data analysis scheme.

The practical value of automated data interpretation rests in the computers' ability to render results in a clear format and to perform statistical operations on these results. Test performance is, thus, documented. "Interpretive" reports may also be supplied to physicians. Just as automation of test operation changes where and how quality control is done, so automation of interpretation also tends to shift the responsibility for data validity and technical interpretation away from the test operator and onto the designer and vendor of the automated technique.

E. The Interplay of Sensitivity, Speed, and Diagnostic Situations

1. Sensitivity

In infectious disease diagnosis, antigen tests are usually applied to body fluid specimens. These specimens may be from patients previously treated with antibiotics or from patients in whom obtaining a specimen with high yield from culture seems unnecessarily difficult or dangerous. Since the concentration of antigens in such specimens is usually quite low, antigen tests must demonstrate impressive chemical sensitivity (detecting nanograms per milliliter) in order to detect a pathogen (at levels of 10^2 to 10^4 CFU/mℓ).[14] When seeking evidence of antigens indicative of an infectious agent in body fluid specimens, physicians act on an implicit assumption that the presence and quantity of antigen in the body fluid correlates with patient morbidity. Before accepting this assumption, one should consider the distinction between *chemical* sensitivity (quantification of the smallest amount of a substance that an assay method can detect) and *clinical* sensitivity (quantification of a test's rate of positivity among patients otherwise defined as having the disease at issue). There is no necessary link between being able to detect smaller amounts of an antigen and being able to diagnose infectious disease correctly. This relation, where is exists, must be demonstrated for particular antigen-disease pairs. It is more likely to hold in the diagnosis of rare or fastidious infectious agents and less likely in diagnosis of infections caused by agents commonly found in a patient's normal flora or environment.[15]

2. Speed

Although physicians tend to think of all antigen tests as rapid, that is, faster than conventional culture techniques, the increased sensitivity of automated tests takes time. Automated methods may be faster than manual techniques at the same degree of chemical sensitivity, but chemically less sensitive manual techniques may be faster and more efficient in a given clinical situation. A diagnostic situation in which this is true at the moment is the rapid detection of bacterial antigens in body fluids.[16] Latex agglutination and coagglutination as manual antigen detection techniques are simpler than competing automated methods. Quality control of the agglutination methods is easier and their performance does not require competence at operating equipment.[17] They can provide results within minutes of a specimen's arrival at the test station and are also easily performed on a sample-by-sample ("STAT") basis. In contrast, competing automated techniques (currently enzyme-linked immunosorbent assays) run in a time frame of hours and are cumbersome and wasteful to use sample-by-sample.

In clinical chemistry, automation dealt successfully with similar problems; in that field miniaturization of equipment and ingenious adaptation of reagents overcame the impasse. Under similar market pressures, it is likely that the "fast, but not fast enough" status of automated bacterial antigen detection methods will also be altered in the interest of more competitive speed. We will consider the influence of this competitive situation on automated method design in greater detail below.

3. Diagnostic Situations that Favor Antigen Detection

There are situations in clinical microbiology in which the rapidity of automated antigen detection tests is already "fast enough". These are diagnostic occasions in which alternative culture-dependent methods for detecting infectious agents are not only slow, but also complicated[18] or dangerous. Diagnostic situations where automated antigen detection techniques are likely to have a speed advantage include those in which they compete with culture techniques to identify anaerobic bacteria, fungi,[19] and mycobacteria. The detection of these three classes of agents also includes instances of the other two indications for antigen detection as well. Successful cultivation and identification of anaerobes is a complicated activity requiring apparatus and specialist expertise. The danger of laboratory infection from culture techniques is a secondary factor in favor of automated antigen detection in mycology. The same relative indication of laboratory danger holds, although to a lesser extent, in mycobacteriology. The culture of mycobacteria is, however, the classic situation for positive culture results being rendered weeks after a patient's death or discharge from hospital.

In other situations, particularly in parasitology[20,21] and virology,[22] it is difficult to estimate the diagnostic efficiency of automated antigen detection compared to culture because it is difficult to define an independent standard against which both culture and antigen detection can be compared. Where the culture techniques themselves provide the standard, antigen detection appears to have clearer advantages in sensitivity over other culture-independent techniques. These techniques include morphologic methods like stool examinations for ova and parasites in parasitology as well as serologic techniques of diagnosis by antibody response, especially in virology. In general, it takes less expertise and experience for a test operator to maintain good unbiased detection performance with an automated antigen detection technique than with a morphologic method. It is especially difficult to maintain a high level of sensitivity in a morphologic method if the prevalence of the organism in question is low. This is precisely the problem in screening stool specimens for ova and parasites in clinical laboratories in the developed world.[23]

In viral diagnosis, antigen detection methods compete with serologic techniques. Two variables usually interpose themselves between a viral infectious agent and serologic diagnosis: time and the immune response itself. Time is needed for the generation of an acute antibody response or between acute and convalescent samples needed to document a titer elevation. The magnitude of the immune response varies from patient to patient. At one extreme, an immune response may even be absent despite infection in some immune-compromised patients; while at the other extreme, polyclonal B-cell activation may blur the response by reproducing nonspecific rises in antibody titers without a specific agent being present. Antigen detection, while maintaining the immunochemical specificity of serological diagnosis eliminates both these middle variables.[24]

The greatest success of automated antigen testing so far has been in dealing with fastidious viruses. Culture of the hepatitis B virus and rotavirus is impractical, even in most reference situations. As described elsewhere in this volume, the hepatitis B virus fortuitously supplies a large source of pure hepatitis B surface antigen. Testing for hepatitis B surface antigen has been integrated into rapid serological diagnosis of that virus to make it sensitive and precise.[25] At the same time, the high sensitivity of antigen detection[26] has permitted reduction of blood transfusion-borne infection through screening for surface antigen in samples of donated blood. The panel format in hepatitis B diagnosis employing different combinations of antigen and antibody tests has shown itself to be extremely flexible. Different groups of tests can be developed according to their different diagnostic purposes (e.g., acute differential diagnosis of jaundice, monitoring of an established hepatitis infection, helping in the decision whether or not to begin steroid therapy, screening candidate populations before administering hepatitis B vaccine, defining successful vaccination, diagnosing gestations at risk for vertical infection, diagnosing vertically infected neonates, or keeping infectious units of blood from being transfused). Compatible automation (where the same indicator-sensor components record the detection of both antigens and antibodies) is clearly the most convenient and economical way to supply panel diagnosis in this variety of contexts.

In rotavirus testing, the automated antigen test competes against a morphologic method, electron microscopy.[27] Electron microscopy is a "gold standard" in rotavirus detection, but very cumbersome to use, requiring an expert operator with considerable interpretive skill.[28] The clinical sensitivity and specificity of the commercially available direct ELISA for rotavirus remains uncertain, perhaps because changes are going on within the "black box" of the method to preabsorb nonspecific reactivity.[29-31] Nevertheless, this automated antigen test has clearly found a place in the diagnostic armamentarium, because it is better than the alternative of waiting for serologies.

F. Bringing the Diagnostic Situation Nearer to the Patient

Automated antigen detection schemes, especially those capable of detecting a variety of antigens may have a major effect not only on *when* but also on *where* rare or fastidious organisms are identified. In the past only reference laboratories could maintain the variety of seldom-used reagents, colonies of test animals, abstruse technical lore, and specialized machinery necessary to diagnose such organisms. The expense and delay in getting a specimen to a place capable of testing it not only led to delayed diagnosis, but also decreased the likelihood of demonstrating the agent in the specimen.

Successful antigen detection provides a diagnosis without cell cultures, odd agars, suckling mice or the expertise needed to handle these media. Automated antigen detection achieves such a diagnosis not only quickly but also often using essentially the same method and usually the same instrument for all antigen (and antibody) testing. This makes diagnosis of rare and fastidious agents available to less specialized laboratories capable of managing automated technique, if the antigen assays are specific enough.

G. Immunochemical Specificity

In the niches in laboratory ecology for which the automated antigen detection methods are competing, chemical specificity (quantification of an assay's ability to distinguish its analyte from interfering substances), is a precondition for successful survival.

Chemical specificity depends mostly on the affinities of the antibodies used to detect the antigen at issue. This is not a characteristic of antigen detection assays upon which automation touches directly, but it is a characteristic vital to the diagnostic success of automated assays. At present there are two sources of antibodies available for use as detector reagents. They are natural antisera raised in animals by antigen challenge and monoclonal antibodies produced by hybridomas. As a reagent source, natural antisera suffer from difficulties in purification and from residual cross-reactivity. There is a residual uncertainty that all the antibodies in a natural serum product are recognizing the same antigen (i.e., that the antibody is "pure") and a residual uncertainty that the antiserum recognizes only the antigen tissue (i.e., the antibody may "cross-react"). Any method using natural immune reagents must control for these uncertainties. This necessity is reflected in complicated method design.

As discussed below in Section IV, automation does play an indirect role, by controlling hybridoma selection, in the provision of the other source of immunochemical specificity. Monoclonal antibodies initially seemed to be the perfect answers to difficulties with natural antisera. Coming from a single progenitor cell, they are genetically pure. Since a clone is selected by a specific antigen, it would seem to follow that if the antigen were pure, then the antibody would be specific. Most antibodies thus selected turned out, however, to have less affinity for the very antigens that selected them than natural antisera achieve.[52] This characteristic often presents a practical trade-off between the high affinity of natural sera and the high specificity of monoclonal antibodies. Raising several different monoclonal antibody "strains" against the same antigen is a tactic that circumvents this trade-off.[56] Such a collection of antibodies may recognize different or overlapping antigenic sites (epitopes) on the same antigen.[32] This more accurate mimic of the natural response than a single monoclonal produces higher (i.e., more "natural") affinity.

A second aspect of the process of obtaining chemical specificity from hybridoma antibodies is antigen presentation. Here also are opposing considerations. On one hand, one wants to screen for monoclonal antibodies with a pure antigen. On the other hand, one wants the antigen to present epitopes to the antibodies that are available in vivo where the antigen may be part of a membrane or bound to a carrier protein. Recombinant DNA sequences can direct the production of pure antigen, but, when cloned antigen is used to screen for hybridoma antibodies, its presentation is not necessarily "natural".[33] Antigens obtained from viable organisms present naturally, but this means that they are set in a context of other possible antigen substances, reintroducing the possibility of cross-reactivity.

In the long run, mapping of the epitopes of the surface of an antigen will eventually solve this quandry. In the meantime, while it is easy to conceive of the sequence (1) clone an antigen, (2) select a cloned antibody, (3) turn the antibody into a specific immunochemical reagent, it is not as easy in practice to carry the sequence to completion (see Section IV below for more detail).

II. THE MOST COMMON FORMAT OF AUTOMATED ANTIGEN TESTING — ELISA[34,35]

We will now consider the consequences of automation for specific automated formats of antigen detection tests. The automatable antigen detection methods that have received the most study and the widest research and clinical application are enzyme immunoassays (EIAs). Although single phase competitive immunoassays have some technical advantages (they are fast and simple) the absence, in the past, of large quantities of labeled antigen has made this kind of assay less practical than immunoassays that incorporate phase separation. Phase separation also permits increased chemical sensitivity.[42] For these reasons, we — like most antigen test designers — will concentrate on phase separation assays.

A. Performance Characteristics of ELISAs[7]

By far the most popular phase separation assay is the enzyme-linked immunosorbent assay (ELISA) in which the products of the linked antigen-antibody and enzyme-substrate reactions are fixed to a solid phase while liquid phase reactants are washed away. Separation steps increase chemical specificity by removing unbound, potentially interfering substances during "washes", but at some cost. The "anchor" antibody coated surface thus has its own characteristics (and problems) of affinity and cross-reactivity.[36] Each surface has a level of specific binding (affinity)[37] above which it will no longer bind the antigen at issue even though antigen is present. Each surface also has a background of nonspecific binding in which substances other than the antigen at issue are attached to the surface. These two factors usually have minor, but measurable, negative effects on the chemical sensitivity and specificity of an ELISA.

1. Phase Separation

The requirement of one or more separation steps (washes) is more influential on test performance. Washes have been found to reduce the chemical sensitivity and reproducibility of the ELISA. In particular, washes are the major negative influence on ELISA reproducibility. This is the underlying reason for the value of automating these separations. If automation can effectively improve the technical reproducibility of these crucial steps, then it can be justified on this basis alone.

2. Advantages

ELISAs have two main advantages that make them attractive candidates for automation aimed at producing antigen detection tests of high sensitivity. First, enzyme-substrate reactions are natural amplifiers of signal. A small number of enzyme labels attached to a detected antigen-antibody complex can convert a large but proportional amount of substrate to a measurable product. This amount of product is, in turn, proportional to the signal received by the sensing device. The combination of the specificity of enzyme for substrate and the natural amplification of enzyme-substrate reactions leads to the second systematic advantage of ELISA: the detection of small numbers of immune complexes by relatively low energy markers. Amplification produces a high level of chemical sensitivity (a good signal) without requiring that the indicator molecule generate by itself a large amount of signal energy. Practically, this means that the indicator reagents are stable, that is, have long shelf lives that reduce reagent costs. This feature is particularly helpful in low-demand tests for rare agents.

3. Signal Generation

Signal generation imposes two systematic constraints on the components of enzyme-substrate indicators. To achieve amplification, the enzyme must have a high turnover rate. To maintain proportionality with the number of detector antigen-antibody complexes, the enzyme must maintain its high turnover tightly bound ("stably conjugated") to the indicator antibody. This tight bond must not interfere with the antibody's specific binding to the antigen at issue.

Largely because of laboratory experience in other contexts, horseradish peroxidase, alkaline phosphatase, and β-galactosidase have been selected from a large group of candidate enzymes as indicators for most ELISAs. For these reagents, enzyme conjugation to immunoglobulin is practical and substrates are available that generate easily measured signals. The availability of widely used indicators, thus, increases the versatility of automated detectors designed to receive the indicators' signals.

4. Signal Detection

A major potential benefit of automated ELISAs is automatic quantification of signal detection. In most early ELISA systems the signals were changes in color, that is, changes in transmission of light in the visible spectrum. Higher energy (therefore more sensitive) signals can be generated by fluorescent (or chemiluminescent)[38,39] or radioactive substrates. These signal detection strategies, which require automated sensors, combine the virtues of enzyme amplification with those of high energy indicators. This combination allows assays to detect signal from less substrate and, therefore, to detect signal faster. Automated sensors thus make possible faster detection of less substrate. High energy indicators depend on this for any clinical advantage speed and small test volume may gain for them.

Clinical chemistry has developed a variety of other techniques beside electromagnetic signal generation to detect enzymatic activity. The most important of these from the clinical laboratory viewpoint is the measurement of changes in optical density, that is, changes in the medium through which a signal of visible light passes. A test designer who applies this sort of measurement to ELISAs must insure that any nonspecific change in the optical density of the medium (interference to the specific signal) is eliminated or controlled. Two sources of such interference are failure to remove the natural activity of the indicator enzyme from a clinical sample and the dissociation of the indicator enzyme from the indicator antibody. Any automated antigen detection assay must explicitly deal with these sources of interference to insure adequate performance.

B. The Relationship of Automated ELISA Design to Performance[40]

In each situation of antigen detection there are intrinsic and extrinsic determinants of a test's format that render it easier or harder to automate. The intrinsic determinants are (1) the antigenic characteristics of the substance to be measured, (2) the antigen recognition characteristics of the immune reagents, (3) the conditions required for adequate signal generation, and (4) the signal reception abilities of the sensing device. The extrinsic determinants are (1) the clinical role of the antigen detection assay (screening, presumptive diagnosis, or confirmation) and (2) the laboratory characteristics of the assay (the number of test requests and samples and the amount of time necessary to turn them into results and reports). These have been discussed above in Section I.

The interaction of intrinsic and extrinsic determinants has, among ELISAs, produced at least four genera of automated formats.

1. Direct ELISAs

In direct ELISAs, antibody to the antigen at issue is first attached to a surface. Such surfaces are usually made of plastic and shaped either as a cup in which the reaction mixture is suspended or as beads suspended in the reaction mixture. This surface is usually prewashed to decrease nonspecific binding (wash #1), then the specimen (which may have been preabsorbed to reduce cross-reactivity or pretreated to release antigen from any native binding in the body fluid) is incubated over the anchor antibody surface (immune reaction #1). This incubation (incubation #1) usually lasts at least an hour and is followed by wash #2. The second incubation is with an enzyme-labeled detector-indicator antibody directed at an epitope on the antigen other than the site bound to the anchor antibody. This incubation is also a matter of hours, and is followed by a third wash. The final operational step is the addition of substrate and the enzyme-mediated generation of signal. In this format the main influences on reproducibility are the three washes (the prewash and the two separation steps). The length of the two incubations plays two desired characteristics against one another: the longer the two incubations, the better the test's sensitivity but the less its claim to being a rapid test. The current turnaround time for such ELISAs is about 5 hr. Critical reagent constraints are, first, whether or not preabsorption or pretreatment of the specimen is necessary to lower cross-reactivity into an acceptable range, and second, how easy it is to bind the indicator enzyme to the detector antibody. The first constraint depends on the amount of background cross-reactivity or native binding in the population of samples tested (an extrinsic characteristic that will vary from population to population). The second constraint influences the test's specificity according to the rate at which the enzyme dissociates from its specific bond and binds nonspecifically to the surface. If the enzyme dissociates and washes away, this will also decrease chemical sensitivity. It influences sensitivity in two other ways: first, by the rate at which the enzyme may obstruct the antibody's active site and, second, by the rate at which antibody obstructs the enzyme's active site. Meeting these requirements — binding the enzyme tightly to the antibody but in a way that does not obstruct either antibody or enzyme active site — has proved to be the main hurdle on the course to production of automated direct ELISAs.

2. Indirect ELISAs

The indirect ELISA can be thought of as a solution to this main reagent problem of the direct method, but a solution achieved at some cost in incubations (time) and washes (reproducibility). In the indirect format, anchor antibody is once again bound to a solid, usually plastic, surface. The assay starts with a prewash (wash #1). There follows an incubation with the clinical specimen (with or without preabsorption or pretreatment of the sample). Wash #2 is followed by an incubation with the detector antibody, but the detector antibody is not bound to any indicator. Instead, after wash #3, a second indicator antibody is added. This antibody is bound to the system's indicator enzyme but its specificity is for the type, class, or usually species of the detector antibody. In this way, the detector and indicator functions have been separated. The practical benefits of this indirect design for automation are, first, that labeled antibody to a type, class, or particularly a species antibody is more likely to be available and well characterized than a labeled antibody against the antigen at issue. Second, one indicator antibody can register the presence of any detector antibody in the type, class or species of its specificity. After the indicator is added, incubation and wash #4 are necessary before substrate is added to generate the test signal.

Regarding the indirect method in general, one must balance negative considerations of (1) increased test run time (the additional incubation and wash usually add at least another hour); (2) decreased reproducibility (four vs. three washes); and (3) increased risk of cross-reactivity against the intrinsic positive consideration of enhanced sensitivity (provided that the detector antibody-indicator antibody reaction is no less avid than the two antibody-antigen reactions) and the extrinsic positive considerations of greater reagent availability, economy, and greater ease of automation. By the last advantage we mean that, with automated methods, ELISAs can assay for a variety of antigens not only on the same instrument but also with the same indicator-signal system. Several assays, therefore, use the same operational protocol. As long as separate controls are included for each anchor/antigen/detector combination, the test operator need learn only one test protocol to perform a variety of antigen detection tests. At the same time, different computer programs can handle the differences among assays in data analysis. In this way automation "generalizes" the indirect ELISA format for antigen detection.

3. Competitive Enzyme Immunoassays[41]

The need for multiple immune reactions is the main drawback of the indirect approach. This disadvantage is avoided by other procedures that employ competitive formats for antigen detection. As mentioned above, these procedures all depend on the availability of labeled antigen. If a quantity of the purified antigen in question is available, and if an enzyme can be bound to this antigen in a way that does not interfere with an anchor antibody's recognition site on the antigen, then a two-wash, one-immune-reaction format is feasible. In such a procedure, the solid surface is prewashed, then the clinical sample and the labeled antigen are added together to compete for the anchor antibodies' binding sites. After wash #2, the substrate is added and a *decrease* of signal in the sample is compared to a control containing only labeled antigen. The decrease in signal is proportional to the amount of antigen in the sample. If the extrinsic condition of antigen availability and the intrinsic condition of no steric hindrance of the critical anchor antibody binding site can be satisfied, this format is both faster and more reproducible than the direct and indirect ELISA formats but systematically less sensitive.[42] Deployment of DNA recombination and cloning may provide a source of large amounts of pure antigen, while fast-reacting or high-energy indicators may permit these potentially automatable competitive EIAs to be performed within the time frame of manual agglutination techniques — minutes rather than hours. Although the competitive format is less sensitive than phase separation, it is probably more sensitive and specific than agglutination techniques.

4. Simultaneous Enzyme Immunoassay[40]

In the pursuit of enhanced reaction speed some automated ELISA designers have even been willing to concede ELISA's main advantage, high sensitivity, in order to compete with less sensitive but faster manual methods. In the simultaneous reaction format, separate incubations and washes are kept to the following minimum: prewashing of the reaction surface, preincubation of some samples with blocking sera, deactivating enzymes (to decrease nonspecific binding and, thus, noise generation), and a wash before adding substrate (that also decreases nonspecific signal). Since the anchor antibody and the detector-indicator antibody must recognize separate, sterically noninterfering sites on the antigen, there is no immunologic reason for not reacting them simultaneously, but problems with a single detector-indicator/antibody-enzyme complex (discussed above regarding the direct sequential ELISA) are also present in the simultaneous format. While the sensitivity of the simultaneous format is less than that of the sequential ELISA format, the critical question is "What is the sensitivity of the simultaneous format compared to the agglutination methods with which they compete for rapidity?" If it is more sensitive than agglutination, then the simultaneous antibody format may, like the competitive format, be a candidate for automated antigen detection in the STAT time frame.

C. Variation on the ELISA Theme

Efforts to deal with one or another source of variation in immunochemical affintiy have led to a variety of automated techniques that are still generically ELISAs. An example of such a variation is in the use of biotin, a low-molecular-weight vitamin, as a label.[41] Using a low-molecular-weight marker decreases steric hindrance produced by combining one macromolecule (the immunoglobulin) with another (most enzymes and all other antibodies). The formation of large molecules and their subsequent polymerization hinders binding to the specific active sites on either or both constituents. Nonspecifically bound macromolecular label, which can lead to nonspecific signal generation, is also harder to remove during reagent preparation or separation than is a low-molecular-weight label. Biotin, besides being small, is easily bound to immunoglobulin. Biotin is also efficiently bound by avidin, an egg white protein. It is to avidin that the indicator enzyme is attached. At the cost of another incubation and separation, the steric distance between the detector immunoglobulin active site and the enzyme active site is thus increased without the addition of the third immune reaction. If the nonspecific binding of avidin is efficiently eliminated (so it binds only with biotin), then a more chemically sensitive and specific assay is produced, decreasing the effect of steric hindrance and increasing immunochemical affinity. This test format retains the operational benefit for automation of a single indicator-signal system serving a variety of detector antibodies.

D. Sample Preparation and ELISA Performance

An important influence on ELISAs (and other antigen assays) is antigen preparation. This part of the first phase of the antigen-detection process varies from assay to assay. This variation depends on the clinical specimen (that is, the degree of background or "noise" in the system) and the availability of antigen to be detected (that is, the strength of the signal). As we mentioned when discussing the production of immune specific reagents, some antigens in their native state are more available to detection than others. Antigen presentation and availability are codeterminants of antigen-antibody affinity, hence they are determinants of assay specificity. The value of various chemical techniques for unmasking the bonding sites of characteristic but hidden antigens or of releasing them from native ligands varies from clinical setting to clinical setting, but preassay sample treatment is another series of incubations and washes amenable to automation.

A particular class of often hidden antigens of infectious agents are enzymes characteristic of the agent but not present in the host or his usual milieu.[34,35] In principle, these "antigens" can be detected by easily automated purely enzymological detection techniques that eliminate the immunoglobulin intermediate factor while still maintaining (enzymatic) specificity of recognition.

Besides antigen preparation that enhances chemical sensitivity by revealing hidden or natively bound antigenic sites, another component of sample preparation is enhancement of chemical specificity with maneuvers that decrease background. Background substances are, in fact, the reason why the washes, despite their lamentable effect on reproducibility, are necessary. Inactivation or removal of nonspecific activity and inhibition is incumbent on any successful single phase EIA. "Disorbing" substances are more common in some specimens (those rich in proteolytic enzymes) than in others. They require dilution or neutralization of their activity if the chemical sensitivity (and specificity) of the assay is to be maintained. This titration is available to automation.

A final problem of sample preparation effects the chemical specificity of immunoassays more than their sensitivities. It is the presence in clinical specimens of substances that react nonspecifically with the anchor or detector antibodies. Three tactics are used in response to these sources of interference. One is to denature the interfering substances so that they lose their immunoglobulin binding capacity. The second tactic is to add another incubation and wash with a "blocking serum" capable of removing the cross-reacting substances without removing the antigen at issue. The final tactic is to denature the anchor antibody, that is, remove the Fc portion of the anchor antibody with which most of the common interacting substances react. All of these ploys clearly make the precedures in which they appear more complex.

E. Automation and Complexity

A major limitation on ELISAs in the clinical laboratory is the number of steps at which critical variation can occur. Efforts to control variation lead to a number of potentially necessary "extra steps" or subroutines in the sample-preparation and product-separation phases of antigen detection. The number and similarity of these incubation and wash routines argues for their operational automation. Dealing with background variation also has analytic consequences. Variation between runs often requires multiple controls on each ELISA run, since each run must generate its own control curve. Automated data analysis, in the form of an instrument-based computer, is a convenient way to deal with this requirement.[42,43a]

III. AUTOMATION OF OTHER ANTIGEN DETECTION SYSTEMS[44a]

A. Immunofluorescence Microscopy

When we consider ELISA antigen detection, we place automated rapid antigen detection by immune reagents in competition with morphologic detection of infectious agents. Immunofluorescence (IF) microscopy, in contrast, combines morphologic recognition and immune specificity. Detection by IF microscopy is achieved through specific binding of a fluorescent-labeled antibody to antigenic determinants on the surface of the organisms at issue.[45] The signal generated is emitted fluorescent light. Its intensity is proportional to factors in three locales: the organism, the immune reagents, and the signal receiver (the microscope). The number of antigenic sites available, the affinity (or affinities) of the fluorochrome-labeled antibody, its concentration in the incubation fluid, the efficiency of the filters along the light path, and the transmission characteristics of the lens in the microscope, all influence the intensity of the fluorescence. Besides signal intensity (vs. nonspecific background fluorescence), localization of the signal within cells or tissues (a morphologic characteristic) can help distinguish signal from noise. Among agents larger than viruses, cellular morphology (cytologic detail) of fluorescent structures can indeed distinguish effectively between signal (true-positive) and noise (false-positive) fluorescence. Commercial fluorescent antibodies (FA) against antigens of a variety of infectious agents are available. These antisera are referred to as direct FA (DFA) conjugates to distinguish them from reagents raised against antibodies to infectious agents, called indirect FA (IFA) conjugates. DFA reagents are available for relatively common agents like *Escherichia coli, Haemophilus influenzae, Klebsiella* spp., *Neisseria gonorrhoeae, N. meningitides, Salmonella* spp., *Shigella* spp., *Staphylococcus* spp., *Streptococcus pneumoniae,* groups A to G streptococci, *Chlamydia trachomatis, Candida albicans,* adenovirus, cytomegalovirus, Epstein-Barr virus (EBV), herpes simplex virus (HSV), influenza virus, and parainfluenza virus. Reagents are also available with specificity to less common and, therefore, harder to diagnose entities: *Bordetella pertussis* and *B. parapertussis, Francisella tularensis, Legionella* spp., *Leptospira* spp., *Pseudomonas pseudomallei, Treponema pallidum,* and rabies, measles, and mumps viruses. Recently, monoclonal antibodies have found a definite place among these reagents with monoclonal preparations performing well in the detection of group A streptococci and *C. trachomatis.* Groups or panels of monoclonals are also able to detect *N. gonorrhoeae* and *Legionella pneumophila* and to distinguish between the type I and type II strains of HSV.

The role of automation in DFA is to separate precisely signal from background intensity and make this distinction reproducible. A person scanning a microscope field for "positive fluorescence" may, in spite of the presence of positive and negative controls, be biased in favor of positivity. This bias would lower the threshold of the intensity registered as positive, especially if few positives were seen. The use of "blind" controls might check this tendency, but, in general, the separation between signal and background would remain implicit and subjective. Automated devices make this distinction explicit and objective by providing a graphic display for positive control, negative control, and sample fluorescent intensities. The pertinent laboratory question is whether or not the introduction of an objective standard would improve the diagnostic performance of the test. The pertinent clinical question arises when the laboratory question is answered in the affirmative: does better test performance lead to better assortment of patients into true-positive and true-negative categories? The answers to these questions await the application of practical registers of fluorescence intensity to large numbers of clinical specimens.

B. Radioimmunoassay[46] and Fluorescence Immunoassay[47]

Phase heterogeneous immunoassays exist that use indicator systems other than signal-generating enzyme-substrate complexes. Indicators of systems that take advantage of radioactivity and fluorescence have been shown to be practical. Phase heterogeneous radioimmunoassays do not have the potential for sensitivity of the self-amplifying ELISAs[42] and are saddled with some practical problems: mischievious reagents, cumbersome sensors, and ponderous operation requiring highly trained operators. At first glance, fluorescence also seems burdened by a major disadvantage. Fluorescence intensity does not seem to be an adequately precise indicator for antigen-detection purposes. This is not true, however, for measurement of fluorescence polarization[10] and fluorescence phase separation.[9] The key technical advance that makes these techniques available to the clinical laboratory is one of automation: the ability to couple the sensing instrument to a computer powerful enough to carry out the sophisticated signal analyses necessary to use these measures as indices. Although not quite as dramatic as the deployment of Fourier transformations to turn indices of nuclear magnetic resonance into images of planes within the human body, the application of mathematical techniques far beyond the ken of test operators and test result interpreters (made possible by the miniaturization of computational power) has widened the application of fluorescence indicators in antigen detection (e.g., pulsed-light time-resolved fluoremetry).[48]

IV. AUTOMATION AND MONOCLONAL REAGENTS IN ANTIGEN DETECTION[49-55]

Because of the importance of immune reagents to the sensitivity of antigen tests, monoclonal antibodies have appeared in a variety of contexts in our discussion of automated antigen detection.[56] Automation can play a major role in the production of these antibodies as well as their application. It appears, in fact, that production and application of monoclonal reagents must be linked for their potential to be realized.

A. Natural and Monoclonal Immune Reagents Contrasted[32,57,58]

To raise titers of natural antibodies, more or less pure antigen is inoculated, usually multiple times, into a set of animals among whom a subset of "good responders" is identified. When the sera from this subset are pooled, two problems appear. First, the same antigen usually produces antisera of various titers, affinities, and specificities not only among different animals in the responder subset, but also from the same animal at different times. Second, cross-reacting and nonspecific antibodies are almost always present in an antiserum. To make a diagnostic reagent, this "natural" variation must be standardized so that the cross-reaction and nonspecificity can be reduced. As briefly alluded to in Section I.G above, these problems make production of immune reagents of defined high specificity difficult. They also make it expensive to produce large quantities of these reagents. Such a variable process is not easily automated. Until the development of hybridomas this cumbersome process was the main technical obstacle to the production of antigen detection reagents.

Hybridomas overcome the problem of heterologous sera; they produce antibodies all of the same specificity, affinity, and titer. Injected into mouse peritoneum, hybridoma cells of the same clone proliferate and produce ascitic fluid containing high titers of the clone's single antibody in large volumes. This production technique decreases significantly the post-production problem of purification. A key step in the hybridoma procedure, the selection of clones with appropriate specificity and adequate affinity may have to be repeated numerous times. This key repetitious step can be automated.

B. The Anatomy of Hybridoma Production[52,59,60]

Hybridomas receive their name because they are the fused product of two cells: a single B-lymphocyte contributes the antibody specificity and a single myeloma cell contributes the abilities of continuous growth in cell culture and tumor production in mice. The specificity and the abilities of two different cell lines are thus hybridized in one cell and its identical descendants. It is in the selection of a desirable clone of descendants that automation can play a role.

The hybridoma process, like natural antibody production, begins with exposure of an antigen to a host animal's immune system. After one or more such immunizations, the host animal's spleen is removed and the spleen cells are mixed with myeloma cells in conditions that first, promote the fusion of cell membranes, and second, select for the survival of spleen-myeloma hybrids. At this point, an assay is necessary to screen quickly but reproducibly for hybrids producing monoclonal antibodies, so that these clones can be selected and cultivated. The semi- to almost fully automated replicating technique supplies this need. In 96-well microtest plates, hybrids are grown to a specific density, automatically determined by a sensing device. Samples of the supernatants of these cultures are then divided among other microtiter plates. The selector antigens have been annealed to the surface of the wells in the second group of plates. Immune specific reactions are detected in positive wells by radio- or enzyme-linked immunoassays. The link between production and application appears with the observation that the clonal selector assays must be similar in signal-receiver format to the assay in which the desired monoclonal(s) will eventually be employed.[52] Monoclonals selected by an ELISA and shown to be sensitive in ELISA detection assays often fail to react in a "less sensitive" fluorescent antibody assay. As discussed above, this may be a matter of signal strength (the enzyme-triggered signal having been amplified by high substrate turnover) or steric hindrance of immune reactive sites by large fluorescent moieties.

After the desired clones are identified, they must be cultivated and reassayed, then reselected until a stable phenotype of antibody production has been established. Only after this repetitious, unpredictable stabilization process (which lasts 4 to 6 months) can the hybridoma cells be moved into their production site in murine peritonea. The advantage of automating the repeated assays in this process is obvious.[61] The development of such screening assays and problems of propagation-stabilization itself are the major technical issues in monoclonal production. Other issues that add to the expense and hazards of producing monoclonal antibodies are the high cost of maintaining so many tissue cultures, the immunogenicity of the antigen at issue in the test animal from whom the splenic B cells are initially harvested (this may be frustratingly different from the same antigen's immunogenicity in human beings or may be too meager to produce cultivatable clones) and finally, even when adequate immunogenicity appears to be documented, the frequency of usable hybrid generation remains unpredictable and low. Since the development of a successful monoclonal immune reagent is seldom a one-shot process, this is another reason for its automation. Nevertheless, by 1984 monoclonal reagents had been raised against at least 15 different bacterial species, 28 different viruses, 8 mycoplasma species and 25 parasites.[62-64]

C. The Chemical Consistency of Monoclonal Antibodies and Automation

The low frequency of generation of usable hybrids is not the only argument for automating large operations in search of specific antibodies. As considered during our discussion of immunochemical specificity above (Section I.G), the relatively high affinity of natural antisera may be due to their containing several clones of antibodies to different or overlapping antigenic sites (epitopes) on the same antigen. Although completely specific, a single monoclonal to the same antigen may prove to have a lower affinity, sensitivity, and even specificity than such a natural mix. A decrease in specificity seems paradoxical, but the monoclonal may be completely specific for its epitope but the epitope itself may be "nonspecific", appearing on surfaces other than that of the organism in question.[65] For these reasons, in some cases of antigen-detection, particularly from clinical specimens, a mixture or "panel" of monoclonals functions more effectively than a single, completely specific antibody.[32,66]

Chemically, monoclonals are, nevertheless, easier reagents to work with than natural antisera. Not only the usual immunochemical properties (complement fixation, precipitation, agglutination), but also physicochemical properties (affinity for enzyme or fluorochrome label, solubility at varying pH, stability at varying temperature) can be much more certainly assessed. Automation in the same microtiter plate format that is used for hybridoma screening and selection can also facilitate the mixing and matching of various candidate monoclonals into an artificial antiserum of maximum affinity and specificity.

D. Monoclonal Antibodies, Antigen Detection, and Flow Cytometry

Monoclonal reagents will appear not only in currently available technologies like ELISA and immunofluorescence techniques, but also in a more sophisticated automated technique which combines the specificity of antigen recognition and the sensitivity of morphologic recognition with exquisite quantitation and analysis: flow cytometry. The flow cytometer has already found clinical application sorting lymphocytes, not only by size, density and nucleic acid content, but also by surface antigens recognized by monoclonals. These same indices can be used to sort out bacterial, mycobacterial, fungal, and parasitic cells as well as mammalian cells, not only by size, shape, and density but also according to their surface antigens.[67,68]

V. SUMMARY AND CONCLUSIONS

Automation can play a role in all phases of the operation of antigen detection procedures as well as in the analysis of the data these procedures produce. Whether automation is feasible and the degree of automation that is appropriate depend on the volume of specimens processed per batch, the importance of test sensitivity in the clinical context, and the turnaround time demanded by the clinical situation. Automation facilitates sensitivity and precision of procedures and tends to shift responsibility for the test's technical performance from the test operator to the manufacturer, as reagents, sensing devices, and interpretive programs in automated format are packaged and marketed as kits. The affinities and absorbances of a procedure's immune reagents are the major determinants of a test's chemical specificity while the interaction of a test's indicator reagents with the sensing device strongly influences its specificity. Through its role in hybridoma technology and by coupling sensitive signal receiving devices to increasingly sophisticated interpretive programs, automation contributes both to the improvement of immune reagent quality and to signal detection performance. Presently, automated antigen detection systems are still slower than manual antigen detection methods, although new test formats are being developed to get automated techniques into the "STAT time frame", while still maintaining automation's relative advantages of sensitivity and precision. In the meantime, automated antigen detection techniques, particularly ELISAs, are playing a larger and larger role as confirmatory and reference tests where precision, sensitivity, and versatility of equipment are important factors.

In conclusion, improved immune reagents (products of molecular biology), sophisticated sensing devices (products of analytical chemistry), and powerful computer programs (products of computer science), have come together to permit the design of automated antigen detection systems. These systems are now seeking their place in the complicated and difficult tasks of infectious disease diagnosis.

ACKNOWLEDGMENT

This work was supported in part by the U.S. Public Health Service Grant No. AI 13150 and No. AI 19094.

REFERENCES

1. Hill, H. R. and Matsen, J. M., Enzyme-linked immunosorbent assay and radioassay in the serologic diagnosis of infectious diseases, *J. Infect. Dis.*, 147, 258, 1983.
2. Arden, D. W., *What Can Be Automated?*, MIT Press, Cambridge, 1980, chaps. 9 and 10.
3. Logan, J. E., Criteria for kit selection in clinical chemistry, in *Clinical Biochemistry: Contemporary Theories and Techniques*, Vol. 1, Spiegel, H., Ed., Academic Press, New York, 1981, 43.
4. Fraser, C. G., Desirable performance standards for clinical chemistry tests, *Adv. Clin. Chem.*, 23, 299, 1983.
5. Schrire, L., Automation in the microbiology laboratory, *Eur. J. Clin. Microbiol.*, 2, 4, 1983.
6. Jatlow, P., Selection of analytic methods for therapeutic drug monitoring, *Human Pathol.*, 15, 404, 1984.
7. Yolken, R. H., Enzyme immunoassays for the detection of infectious antigens in body fluids: current limitations and future prospects, *Rev. Infect. Dis.*, 4, 35, 1982.
8. Landon, J. and Kamel, R. S., Immunoassays employing reactants labelled with a fluorophore, in *Immunoassays for the 80s*, Voller, A., Bartlett, A., and Bidwell, D., Eds., University Park Press, Baltimore, 1981, chap. 3.
9. McGown, L. B. and Bright, F. V., Phase-resolved fluorescence spectroscopy, *Anal. Chem.*, 56, 1400A, 1984.
10. Dandliker, W. B., Hsu, M.-L., Levin, J., and Rao, B. R., Equilibrium and kinetic inhibition assays based upon fluorescence polarization, *Methods Enzymol.*, 74, 3, 1981.
11. Hicks, J. M., Fluorescence immunoassay, *Human Pathol.*, 15, 112, 1984.
12. McLafferty, F. W., Trends in analytical instrumentation, *Science*, 226, 251, 1984.
13. Michael, B. S. and Davis, B. G., Automation in immunology, *J. Med. Technol.*, 2, 213, 1985.
14. Salzman, G. C. and Gregg, C. T., Current and experimental methods of rapid microbial identification, *Bio/Technol.*, 2, 143, 1984.
15. Kerkering, T. M., Diagnosing invasive candidiasis: the clinician's albatross, *Clin. Microbiol. Newsl.*, 4, 25, 1982.
16. Fung, J. C. and Tilton, R. C., Detection of bacterial antigens by counterimmunoelectrophoresis, coagglutination, and latex agglutination, in *Manual of Clinical Microbiology*, 4th ed., Lennette, E. H., Balows, A., Hausler, W. J., Jr., and Shadomy, H. J., American Society for Microbiology, Washington, D.C., 1985, chap. 90.
17. Plorde, J. J., Newer methods in microbial diagnosis, in *Septic Shock*, Sande, M. A. and Roote, R. K., Eds., Churchill Livingstone, Edinburgh, 1985, chap. 10.
18. Miettinen, A., Turunen, H., Paavonen, J., Jansson, E., and Leinikki, P., Detection of *Mycoplasma hominis* antigen in clinical specimens by using a four layer modification of enzyme immunoassay (EIA), *J. Immunol. Methods*, 69, 267, 1984.
19. Kaufman, L. and Reiss, E., Serodiagnosis of fungal diseases, in *Manual of Clinical Microbiology*, 4th ed., Lennette, E. H., Ballows, A., Hausler, W. J., Jr., and Shadomy, H. J., American Society for Microbiology, Washington, D.C., 1985, chap. 96.
20. Walls, K. W., Serodiagnostic tests for parasitic diseases, in *Manual of Clinical Microbiology*, 4th ed., Lennette, E. H., Balows, A., Hausler, W. J., Jr., and Shadomy, H. J., American Society for Microbiology, Washington, D.C., 1985, chap. 97.
21. McCabe, R. E. and Remington, J. S., The diagnosis and treatment of toxoplasmosis, *Eur. J. Clin. Microbiol.*, 2, 95, 1983.
22. Kurstak, E., Tijssen, P., and Kurstak, C., Enzyme immunoassays applied in virology, in *Applied Virology*, Kurstak, E., Ed., Academic Press, Orlando, Fla., 1984, chap. 29.
23. Ungar, B. L. P., Yolken, R. H., Nash, T. E., and Quinn, T. C., Enzyme-linked immunosorbent assay for the detection of *Giardia lamblia* in fecal specimens, *J. Infect. Dis.*, 149, 90, 1984.
24. Aranjo, F. G. and Remington, J. S., Antigenemia in recently acquired acute toxoplasmosis, *J. Infect. Dis.*, 141, 144, 1980.
25. Govindarajan, S., Asheavai, M., Chau, K. H., Nevalainen, D. E., and Peters, R. L., Evaluation of enzyme immunoassay for anti-HBc IgM in the diagnosis of acute hepatits B virus infection, *Am. J. Clin. Pathol.*, 82, 232, 1984.
26. Ben-Porath, E., Wands, J., Gruia, M., and Isselbacker, K., Clinical significance of enhanced detection of HBsAg by a monoclonal radioimmunoassay, *Hepatology*, 4, 803, 1984.
27. Selb, B., Baumeister, H. G., Maass, G., and Doerr, H. W., Detection of human rotavirus by nucleic acid analysis in comparison to enzyme-linked immunoassay and electron microscopy, *Eur. J. Clin. Microbiol.*, 4, 41, 1985.
28. Brandt, C. D., Kim, H. W., Rodriguez, W. J., Thomas, L., Yolken, R. H., Arrobio, J. O., Kapikian, A. Z., Parrott, R. H., and Chanock, R. M., Comparison of direct electron microscopy, immune electron microscopy, and rotavirus enzyme-linked immunosorbent assay for detection of gastroenteritis viruses in children, *J. Clin. Microbiol.*, 13, 976, 1981.

29. Krause, P. J., Hyams, J. S., Middleton, P. J., Herson, V. C., and Flores, J., Unreliability of rotazyme ELISA test in neonates, *J. Pediatr.*, 103, 259, 1983.
30. Cukor, G., Perron, D. M., Hudson, R., and Blacklow, N. R., Detection of rotavirus in human stools by using monoclonal antibody, *J. Clin. Microbiol.*, 19, 888, 1984.
31. Rotbart, H. A., Yolken, R. H., Davis, D., Roe, M. H., Nelson, W. L., and Levin, M. J., Validation of Rotazyme ELISA Test Results in Newborns, presented at 23rd Annu. Interscience Conf. on Antimicrobiol. Agents and Chemotherapy, Washington, D.C., October 8—10, 1984, Abstr. 776.
32. Nowinski, R. C., Tam, M. R., Goldstein, L. C., Stong, L., Kuo, C. C., Corey, L., Stamm, W. E., Handsfield, H. H., Knapp, J. S., and Holmes, K. K., Monoclonal antibodies for diagnosis of infectious diseases in humans, *Science*, 219, 637, 1983.
33. Kirchhoff, L. V., Hieny, S., Shiver, G. M., Shary, D., and Sher, A., Cryptic epitope explains the failure of a monoclonal antibody to bind certain isolates of *Trypanosome cruzi*, *J. Immunol.*, 133, 2731, 1984.
34. Josephson, S. L., Immunoenzyme techniques, *Clin. Microbiol. Newsl.*, 7, 81, 1985.
35. Yolken, R. H., Enzyme analysis for the diagnosis of infectious diseases, *Ann. N.Y. Acad. Sci.*, 428, 223, 1984.
36. Shekarchi, I. C., Sever, J. L., Lee, Y. J., Castellano, G., and Madden, D. L., Evaluation of various plastic microtiter plates with measles, toxoplasma, and gammaglobulin antigens in enzyme-linked immunosorbent assays, *J. Clin. Microbiol.*, 19, 89, 1984.
37. Rotmans, J. P. and Scheven, B. A. A., The effect of antigen cross-linking on the sensitivity of the enzyme-linked immunosorbent assay, *J. Immunol. Methods*, 70, 53, 1984.
38. Boeckx, R. L., Chemiluminescence: applications for the clinical laboratory, *Human Pathol.*, 15, 104, 1984.
39. Woodhead, J. S., Sturgess, M., Jones, M. K., and Weeks, I., Chemiluminescence labels in immunoassay, in *Monoclonal Antibodies and New Trends in Immunoassay*, Bizollon, C. A., Ed., Elsevier, New York, 1984, 165.
40. Yolken, R. H., Solid-phase enzyme immunoassays for the detection of microbial antigens in body fluids, in *Manual of Clinical Microbiology*, 4th ed., Lennette, E. H., Balows, A., Hausler, W. J., Jr., and Shadomy, H. J., American Society for Microbiology, Washington, D.C., 1985, chap. 98.
41. Kabakoff, D. S. and Greenwood, H. M., Recent developments in homogeneous enzyme immunoassay, in *Recent Advances in Clinical Biochemistry 2*, Alberti, K. G. M. M. and Price, C. P., Eds., Churchill Livingstone, Edinburgh, 1981, chap. 1.
42. Ekins, R. and Jackson, T., Non-isotopic immunoassay — an overview in *Monoclonal Antibodies and New Trends in Immunoassays*, Bizollen, C. A., Ed., Elsevier, New York, 1984, 149.
43. Wilchek, M. and Bayer, E. A., The avidin-biotin complex in immunology, *Immunol. Today*, 5, 39, 1984.
43a. Slezak, T. R., Vanderlaan, M., and Jensen, R. H., A computer-based data analysis system for enzyme-linked immunosorbent assays, *J. Immunol. Methods*, 65, 83, 1983.
44. Boitieux, J. L., Romette, J. L., Aubry, N., and Thomas, D., A computerized enzyme immunosensor: application for the determination of antigens, *Clin. Chem. Acta*, 136, 19, 1984.
44a. Kricka, L. J., Stanley, P. E., Thorpe, G. H. G., and Whitehead, T. P., *Analytical Applications of Bioluminescence and Chemiluminescence*, Academic Press, London, 1984.
45. McKenny, R. M. and Cherry, W. B., Immunofluorescence microscopy, in *Manual of Clinical Microbiology*, 4th ed., Lennette, E. H., Balows, A., Hausler, W. J., Jr., and Shadomy, H. J., American Society for Microbiology, Washington, D.C., chap. 9.
46. Witherspon, L. R., Immunoassay — is there a future role for nuclear medicine?, *J. Nucl. Med.*, 24, 952, 1983.
47. Schall, R. F. and Tenoso, J. H., Alternative to radioimmunoassay: labels and methods, *Clin. Chem.*, 27, 1157, 1981.
48. Soini, C., Pulsed light, time-resolved fluorometric immunoassay, in *Monoclonal Antibodies and New Trends in Immunoassays*, Bizollen, C. A., Ed., Elsevier, New York, 1984, 197.
49. Cone, R., Application of monoclonal antibodies to clinical diagnostic assays, *J. Med. Technol.*, 2, 279, 1985.
50. Goding, J. W., *Monoclonal Antibodies: Principles and Practice*, Academic Press, London, 1983.
51. Ferrone, S. and Dierich, M. P., *Handbook of Monoclonal Antibodies: Application in Biology and Medicine*, Noyes Publications, Park Ridge, N.J., 1985.
52. Campbell, A. M., Monoclonal antibody technology, in *Laboratory Techniques in Biochemistry and Molecular Biology*, Vol. 13, Burdon, R. H. and von Knippenberg, P. H., Eds., Elsevier, Amsterdam, 1984.
53. Edwards, P. A. W., Some properties and applications of monoclonal antibodies, *Biochem. J.*, 200, 1, 1981.
54. Yelton, D. E. and Scharff, M. D., Monoclonal antibodies: a powerful new tool in biology and medicine, *Annu. Rev. Biochem.*, 50, 657, 1981.

55. Pollock, R. R., Teillaud, J. L., and Scharff, M. D., Monoclonal antibodies: a powerful tool for selecting and analyzing mutations in antigens and antibodies, *Annu. Rev. Microbiol.*, 38, 389, 1984.
56. Krakauer, H., Clinical applications of monoclonal antibodies, *Eur. J. Clin. Microbiol.*, 4, 1, 1985.
57. Mitcheson, D. A. and Coates, A. R. M., Monoclonal antibodies in bacteriology in *Monoclonal Antibodies in Clinical Medicine*, McMichael, A. J. and Fabre, J. W., Eds., Academic Press, London, 1982, chap. 12.
58. Tung, A. S., Production of large amounts of antibodies, nonspecific immunoglobulins and other serum proteins in ascitic fluids of individual mice and guinea pigs, *Methods Enzymol.*, 93, 12, 1983.
59. Largone, J. J. and Vunakis, H. V., Eds., Immunochemical techniques Part E: monoclonal antibodies and general immunoassay methods, in *Methods in Enzymology*, Vol. 93, Colowick, S. P. and Kaplan, N. O., Eds., Academic Press, New York, 1983.
60. Kohler, G., Derivation and uses of monoclonal antibodies, in *From Genetic Engineering to Biotechnology — The Critical Transition*, Whelan, W. J. and Black, S., Eds., John Wiley & Sons, New York, 1982, 107.
61. Jolley, M. E., Warz, C.-H. J., Ekenberg, S. J., Zuelke, M. S., and Kelso, D. M., Particle concentration fluorescence immunoassay (PCFIA): a new, rapid immunoassay technique with high sensitivity, *J. Immunol. Methods*, 67, 21, 1984.
62. Polin, R. A., Monoclonal antibodies against microorganisms, *Eur. J. Clin. Microbiol.*, 3, 387, 1984.
63. Porterfield, J. S. and Tobin, J. O'H., Viral and bacterial infectious diseases, *Br. Med. Bull.*, 40, 283, 1984.
64. Cohen, S., Monoclonal antibodies in parasitic infectious diseases, *Br. Med. Bull.*, 40, 291, 1984.
65. Garson, J. A., Beverley, P. C. L., Coakham, H. B., and Harper, E. I., Monoclonal antibodies against human T lymphocytes label Purkinje neurones of many species, *Nature (London)*, 298, 375, 1982.
66. Ehrlick, P. H., Moyle, W. R., Moustafa, Z. A., and Canfield, R. E., Mixing two monoclonal antibodies yields enhanced affinity for antigen, *J. Immunol.*, 128, 2709, 1982.
67. Sahar, E., Lamed, R., and Ofek, I., Rapid identification of *Streptococcus pyogenes* by flow cytometry, *Eur. J. Clin. Microbiol.*, 2, 192, 1983.
68. Van Dilla, M. A., Langlois, R. G., Pinkel, D., Yajko, D., and Hadley, W. K., Bacterial characterization by flow cytometry, *Science*, 220, 620, 1983.

Chapter 10

IMMUNOASSAYS FOR MEASUREMENT OF ANTIMICROBIAL AGENTS IN BODY FLUIDS

Michael A. Pfaller

TABLE OF CONTENTS

I. INTRODUCTION

One of the more striking changes in the field of laboratory medicine in the last 10 to 15 years has been the increased demand for therapeutic drug monitoring (TDM). This is true not only for agents such as the anticonvulsants and the cardioactive drugs but also for certain antimicrobial agents such as the aminoglycosides and vancomycin. In large part this increase can be attributed to the development of immunoassays, particularly the nonisotopic homogeneous immunoassays. These assays are rapid, safe, and simple to perform and have made many traditional drug assay systems, including the microbiologic bioassays, obsolete for many laboratories. The combination of these technical advances and increased demand for quantitation of antimicrobial agents has placed many microbiology laboratories in a position of being forced to choose between several types of assay systems. The purpose of this chapter is to discuss the commercially available immunologic assays for the quantitation of aminoglycosides and vancomycin, the two most heavily monitored groups of antimicrobial agents, in terms of technology, comparative studies, cost-effectiveness, and clinical indications for monitoring. In this way I hope to provide information which will be useful for the microbiologist faced with deciding which of the many available systems would be best suited for use in the microbiology laboratory.

II. THERAPEUTIC DRUG MONITORING OF ANTIMICROBIAL AGENTS

A. Rationale

The number of drugs which have been rigorously shown to be worth monitoring clinically as a guide to therapy is extremely small.[1-3] In general, therapeutic drug monitoring is performed in order to check compliance and ensure that the dosage program for an individual patient has resulted in therapeutically effective levels which are unlikely to cause toxic effects.[1-3] This rationale is based on three general assumptions as follows:

1. The therapeutic response correlates better with the blood level of the agent than with the amount given.
2. An optimum therapeutic range has been established such that most patients receive maximum clinical benefit and minimum toxicity.
3. Adjustment of the drug dosage to achieve the therapeutic range, as measured by laboratory assays, results in a more favorable therapeutic outcome than would be achieved by the exercise of sound clinical judgement alone.

In the past several years the validity of each of these assumptions has been established for the aminoglycosides and to some extent for vancomycin. It is clear that a very low toxic to therapeutic ratio exists for the aminoglycosides and that excessively high peaks and/or trough levels are associated with significant nephro- and ototoxicity.[4,5] Marked variations in serum levels have been demonstrated in patients receiving similar dosages of these agents.[6,7] Clinical studies have demonstrated the superiority of quantitation of serum levels of aminoglycoside vs. the use of a nomogram in determining dosage requirements.[6,8] Finally, a therapeutic range has been established for these agents with demonstration of improved outcome when blood levels are maintained within this range.[9-11] Although vancomycin has a much greater therapeutic ratio than the aminoglycosides there is clearly ototoxicity associated with elevated serum levels[12,13] as well as extreme variation in serum levels from patient to patient on similar doses of the drug.[12-16] Thus, there is a need to monitor serum levels of vancomycin to avoid toxicity and to assure adequate serum levels for optimal therapy.

B. Clinical Indications

Given the above rationale for therapeutic drug monitoring, one must consider the clinical indications for monitoring the blood (or other body fluid) levels of antimicrobial agents. Such guidelines are especially important to keep in mind and reemphasize to the ordering physician, in light of several recent studies suggesting that as few as 20% of all aminoglycoside levels are ordered and/or interpreted appropriately.[17,18] This is an obvious example of the type of misuse of laboratory testing that must be vigorously attacked by both the laboratorian and the clinician. Such misuse of laboratory testing could not only lead to suboptimal therapeutic outcome, but, in addition, to increased costs to both patients and the institution. Although the technological advances in the area of TDM have made serum aminoglycoside and vancomycin levels extremely easy to perform we must always consider the cost, as well as quality, of sample analysis when deciding whether or not to perform these assays. The most efficient utilization of health care dollars can only be realized with appropriate sample collection, and appropriate patient selection for drug level determinations.[17-21]

The clinical indications for monitoring serum levels of antimicrobial agents are generally similar to those outlined for many other pharmacologic agents.[1,2] Studies of the pharmacology and clinical efficacy of both aminoglycosides and vancomycin provide the basis for specific recommendations for appropriate patient selection for serum determinations of these agents.[3,7,11,21] These recommendations are as follows:[3,21]

1. Patients with serious infections in whom it is essential to ensure adequate serum concentrations early in the course of therapy[9-11]
2. Patients whose renal function is changing rapidly or unpredictably and in whom it is not possible to predict accurately the concentrations achieved after several doses (renal disease, renal transplant, or dialysis patients[8,13-15,22])
3. Patients in whom dosage may correlate poorly with serum concentrations (burn or trauma patients,[6-9] patients with edema, ascites, or morbid obesity[16,23,24])
4. Patients who are receiving a prolonged course of therapy[25]
5. Patients who have received previous courses of aminoglycosides[25] or are receiving other drugs which may potentiate aminoglycoside or vancomycin toxicity (diuretics,[26] amphotericin B[27])
6. Patients receiving these agents for an infection with a susceptible organism and showing little or no clinical response[9-11,21]

Clearly these are only general guidelines and there most certainly will be exceptions and additions to those listed. However, for the reasons mentioned previously, efforts should be made by both the laboratory and the clinician to establish these or similar guidelines in order to maximize the useful information obtained from the monitoring of serum levels of antimicrobial agents and at the same time minimize cost and confusion.

C. Requirements for a Clinical Assay System

In considering the newer semiautomated and automated instrumentation currently available it is useful to keep in mind some general requirements that should be fulfilled by an assay system in order to provide an optimal therapeutic drug monitoring service.[1,2] In order to be practically useful in a routine clinical laboratory an assay should (1) be specific, such that it detects the antimicrobial agent of choice yet is free from interference by other drugs and substances that may be present in the blood of patients, (2) provide a precision and accuracy appropriate for clinical purposes (somewhat arbitrarily designated as a coefficient of variation (CV) of 10 to 15% and accurate to within 20% of the true value[2,3,28,29]), (3) be simple enough to allow performance by personnel with a wide range of expertise both during and outside of normal working hours, (4) yield the results before the next dose will be given and provide "STAT" capability when necessary, (5) require a minimum of skilled technician time, (6) be inexpensive with regard to reagents and instrumentation, and (7) have a sensitivity that enables the use of small sample volumes of patient serum.

III. IMMUNOASSAY SYSTEMS

A. General Considerations

All immunoassays depend upon the noncovalent, reversible binding of the antigen (antimicrobial agent) by specific antibodies. This reaction, which obeys the law of mass action, may be described by the following equation:

$$Ag + Ab \underset{k_2}{\overset{k_1}{\rightleftharpoons}} Ag{:}Ab$$

where Ag represents the antigen (antimicrobial agent), and Ab the specific antibody, both of which are free in solution. The antigen-antibody complex, or bound fraction, is represented by Ag:Ab. The rate constants k_1 and k_2 represent the rates of combination and dissociation respectively. The reaction is normally allowed to proceed until equilibrium has been attained. Under these conditions the amounts of antigen in the free and bound fractions will remain constant.

In the immunoassay of an antibiotic the principles of competitive binding are utilized.[1-3] The samples (and standards) are incubated with a small, constant amount of the antibiotic tagged with a specific label (radioisotope, enzyme, fluorochrome) and a constant amount of antiserum. The dilution of antiserum has been selected such that even in the absence of (unlabeled) antibiotic in a sample, not all the labeled antigen will be bound. After equilibrium has been reached, depending upon the configuration of the assay system and the label involved, a measurement of the labeled antigen in the bound or the free (or both) fractions is made. The labeled antigen functions as a tracer to enable the percentage of the total antigen present in the bound fraction to be determined. By using a series of unlabeled standards a standard curve is constructed and the concentration of antigen present in clinical specimens is ascertained.

If the label employed is an isotope (radioimmunoassay, RIA) then the bound and free fractions must be separated using one of a variety of physical, chemical, or immunological techniques prior to determining the radioactive counts in the appropriate fraction (heterogeneous assay). Likewise, many assay systems utilizing fluorescent labels (fluorescence immunoassay, FIA) require a separation step.[30,31] Most of the newer immunoassay systems currently available employ a nonisotopic label, usually an enzyme or a fluorescent group.[3,30-35] The most significant advantage to the use of nonisotopic label is that it permits the development of a homogeneous assay system. These systems take advantage of the change in enzymatic or fluorescence activity produced by the binding of antibody to the labeled antigen to produce an assay which is free of the separation step required of RIA and many FIA procedures.[3,31-33] This has resulted in a number of rapid, simple assays for the quantitation of antimicrobial agents.

To date immunoassay techniques for antimicrobial agents have been limited to the aminoglycosides and vancomycin. Both heterogenous (RIA, FIA) and homogeneous assays, using a variety of labels, are commercially available for the quantitation of these agents. The principles of each of the commonly available assay systems will be discussed in detail in the following sections.

B. Isotopic Immunoassay

Radioimmunoassay (RIA) has been well established as a sensitive procedure for quantitating many different compounds.[2,3,32] The RIA method consists of a labeled antimicrobial agent, a specific antibody, and an unlabeled antimicrobial agent (patients serum and/or standard). An instrument to quantitate radioactivity (usually a gamma counter) is also required. The label employed in all commercially available assays is ^{125}I. Both the labeled and unlabeled antimicrobial agent compete for binding sites on the antibody. If high concentrations of the agent of interest are present in patients' serum, they bind the antibody sites, leaving unbound labeled drug. Free and bound drug are separated prior to counting in a scintillation counter by one of several physical, chemical or immunologic techniques.[2,3,32] The amount of labeled drug bound to the antibody is inversely proportional to the concentration of antimicrobial agent in the patients serum. All of the commonly used aminoglycosides and vancomycin are readily quantitated with commercially available RIA procedures from a variety of manufacturers.[32]

C. Nonisotopic Immunoassay

1. Assays Measuring Antigen-Antibody Complex Formation Rate Nephelometric Inhibition Immunoassay (RNI)

When light of an appropriate wavelength strikes a particle, such as an antigen-antibody complex, some light is reflected, some is scattered, some absorbed and some transmitted.[32,35] The analytical technique of nephelometry examines the light that is scattered at a variety of angles and measures a small increment of signal on a supposedly black or null background.[32,35] Nephelometric analysis can utilize either a static end-point approach (blank value vs. reading at $t = x$) or a kinetic approach (measures the rate of increase in light scatter). The kinetic approach (rate nephelometry) is much faster than the end-point approach and has been useful in the development of several new immunoassays.

The rate nephelometric inhibition assay for aminoglycosides (gentamicin, tobramycin) developed by workers at Beckman (Beckman ICS, Beckman Instruments, Inc., St. Louis, Mo.) utilizes the fact that free aminoglycoside bound to its specific antibody does not produce the degree of light scattering seen with larger immune complexes in which the same aminoglycoside has previously been chemically linked to apoferritin. Thus, an assay has been developed in which the free aminoglycoside in a serum sample competes with an aminoglycoside-apoferritin complex for antiserum binding sites.[32,36] The presence of free drug in the serum results in a decrease in light scattering. The method has been applied to basic rate nephelometric instrumentation in which the diluted sample and reagents are delivered to a reaction cell. The peak rate occurs in less than 60 sec and the measured response for each sample is adjusted by a calibration factor based on testing a single level standard (4.2 μg/mℓ for gentamicin and tobramycin) prior to derivation of the concentration from the calibration curve.[32,36,37] The technique does not require running a complete standard curve for calibration as all reagents supplied by the manufacturer have been previously calibrated and the data concerning the characteristics of the reagent are supplied with each vial of antiserum in the form of a microprocessor card.[36] This data is entered into the instrument by inserting the program card and then requires a single point calibration to be carried out by the operator to standardize the instrument. The system comes in a fully automated version that functions as a discrete analyzer allowing quantitation of several different proteins and drugs on a single patient specimen.

2. Fluorescence Immunoassay (FIA)

a. Assays Requiring a Separation Step (Heterogeneous)

By substitution of a fluorescent dye for the radioactive label, assays similar to commonly used radioimmunoassays can be used for quantitation of aminoglycosides and vancomycin in serum. In a heterogeneous FIA the fluorescent label functions in a manner analogous to a radioactive label, i.e., the activity of the fluorophore is unaffected by its attachment to the antibody and, as with RIA, this type of FIA requires a separation step.[2,3,30-32]

Because the only fundamental difference between FIA and RIA is the label employed, it follows that most of the steps involved in performing the RIA (described above) are also involved in the FIA. The only difference between these two heterogeneous assays is the method of end-point detection.[30-32]

As with the RIA, a variety of chemical, physical, and immunologic techniques have been employed in the separation step required in heterogeneous FIA. The separation step has two important functions in this type of assay. First, to separate the bound and free fractions and, second, to allow removal of any endogenous fluorophores or other interfering substances present in the sample prior to end-point detection.[30-32] This second function enables the assay of larger serum volumes and increases the sensitivity of the assay provided there are adequate wash steps.

b. Assays Which do not Require a Separation Step (Homogeneous)

i. Substrate Labeled Immunofluorescence Assay (SLIFA)

The SLIFA for aminoglycosides (amikacin, gentamicin, tobramycin) was originally described by Burd and co-workers[40] and is commercially available as the Ames TDA® System (Ames Division, Miles Laboratories, Inc., Elkhart, Ind.). The assay is based on a fluorgenic substrate-analyte conjugate in which the drug is covalently attached to a derivative of the fluorogenic enzyme substrate, umbelliferyl-β-D-galactoside. This fluorogenic drug reagent (FDR) is nonfluorescent under the conditions of the assay; however, hydrolysis catalyzed by β-galactosidase yields a fluorescent product.[30,32,33,40] When the antibody to the drug reacts with the FDR, the complex is virtually inactive as a substrate for β-galactosidase. Thus, antibody-bound FDR and free FDR can readily be distinguished since only the free FDR will function as a substrate for the enzyme. These features of the FDR, reactivity with β-galactosidase, and antibody inhibition of that activity, have made it possible to develop a homogeneous competitive-protein binding assay to measure the level of aminoglycosides in serum.[33,40]

The assay of as many as 20 specimens can be performed within 1 hr using a manual system. This assay has also been adapted to a semiautomated instrument (Fluorostat), as well as a variety of automated centrifugal analyzers.[32] This has resulted in a significant improvement in the speed, simplicity, and precision of the assay.[30,32]

ii. Fluorescence Polarization Immunoassay (FPIA)

The principle of FPIA has been known for several decades,[34,41] but because of the lack of suitable instrumentation, the technique has not been used clinically until the recent introduction of the TDX® system (Abbott Laboratories, Irving, Tex.). FPIA is based on the fact that when plane polarized light falls on an antigen labeled with a fluorophore, molecules with their long axes parallel to the plane of light are preferentially excited. If they also rotate between excitation and emission the resultant fluorescence is nonpolarized. The degree of depolarization depends on the rate of rotation which, in turn, is inversely proportional to the size of the molecule. The binding of a low-molecular-weight labeled antigen to antibody produces a large complex which, unlike the native antigen, is only capable of rotating slowly so that much of the emitted fluorescence is still polarized.[34,41]

The principal reagents for FPIA are fluorescent-labeled antigen (tracer) and antibodies specific for the antigen. The measurement of fluorescence polarization provides a direct measure of bound and free tracer in a homogeneous, competitive binding immunoassay. The extent of the depolarization of the emitted fluorescence is proportional to the concentration of unlabeled antigen originally present in the solution.[34,41] The precise relationship between the concentration of unlabeled antigen and polarization is established by generating a standard curve.

The Abbott TDX® system uses FPIA to quantitate the aminoglycosides (amikacin, gentamicin, tobramycin, and netilmicin) and vancomycin in serum and other body fluids. The system utilizes a totally automated, bench-top, fluorescence polarization analyzer, in which all processing is carried out in a light-tight, temperature-controlled environment.[39,41-44] All dilution, pipetting, incubation, and reading steps are automatically performed within the instrument.[39,41-44] The standard curve, which is stored in memory, is stable for up to 4 weeks. A single assay can be performed in approximately 5 min and up to 80 specimens can be analyzed in 1 hr.[39,41-45]

3. Homogeneous Enzyme Immunoassay

The only homogeneous enzyme immunoassay commercially available at present is the EMIT (Enzyme Multiplied Immunoassay Technique) system (Syva, Palo Alto, Calif.). The EMIT methodology was first introduced by Syva in 1974 and has been used extensively for the past several years for the determination of serum levels of aminoglycosides.[38,46-49] The principle of the EMIT procedure involves competitive binding between aminoglycoside in the sample and a known amount of enzyme-labeled aminoglycoside for an anti-aminoglycoside antibody. The enzyme employed is glucose 6-phosphate dehydrogenase. Binding of the enzyme-labeled aminoglycoside decreases the activity of the enzyme. This type of antibody-mediated enzyme inhibition was first demonstrated by Rubenstein and co-workers[50] and is thought to take place either by sterically preventing access of the substrate to the catalytic site or by inducing conformational changes in the enzyme which prevents catalysis.[50,51] The remaining enzymatic activity results in conversion of the coenzyme nicotinamide adenine dinucleotide (NAD) to reduced nicotinamide adenine dinucleotide (NADH). The resulting absorbance change is measured spectrophotometrically. Because each enzyme molecule is capable of catalyzing the conversion of an extremely large number of substrate molecules from an optically inert to optically absorbent substance the EMIT system is at least theoretically extremely sensitive.[50]

The amount of enzyme activity remaining in the test system is directly proportional to amount of free drug originally present in the sample. The serum drug concentration is determined by comparing the enzyme activity with that produced by a series of standard solutions of the same drug. All calculations and data manipulation are performed by a microprocessor available with the system.

The system is available in a manual format including a reagent package designed for "STAT" processing of single specimens (EMIT-Qst). A variety of semiautomated and automated applications are also available to facilitate handling of larger numbers of specimens and to accomodate other EMIT testing. This includes several centrifugal enzyme analyzers typically used in clinical chemistry laboratories. Adaptation of EMIT methods to these analyzers markedly decreases the cost per test in addition to providing excellent precision, speed, and simplicity.[52-54] The Syva Advance system is an immu-

Table 1
FEATURES OF SIX AMINOGLYCOSIDE IMMUNOASSAYS

Test method[a]	Detection method	Antigen label	Separation step	Standard curve stability	Detection range[b] ($\mu g/m\ell$)	Instrumentation
RIA[c]	Isotope	^{125}I	Yes	Each run	0.1—16	Gamma counter
RNI (Beckman)	Nephelometry	None	No	Calibrators only — each run	0.5—16	Nephelometer
FIA[c]	Fluorescence	Fluorescein	Yes	Each run	0.5—16	Fluorometer
SLIFA (Ames)	Fluorescence	β-Galactosyl-umbelliferone	No	Each run	1—12	Fluorometer
FPIA (Abbott)	Fluorescence	Fluorescein	No	2—4 weeks	0.2—10	Fluorescence polarization analyzer
EMIT (Syva)	Absorption	G-6-PDH	No	24 hr	1—16	Spectrophotometer pipettor-diluter

[a] Abbreviations are as follows: RIA, radioimmunoassay; RNI, rate nephelometric inhibition; FIA, fluorescent immunoassay; SLIFA, substrate-labeled immunofluorescent assay; FPIA, fluorescence polarization immunoassay. The term EMIT is a registered trademark of Syva Co., Palo Alto, Calif.

[b] The detection range listed in each case is for the determination of *gentamicin*.

[c] The kits for RIA and FIA are available from a variety of manufacturers.

nochemistry system which utilizes the EMIT principle and detects the end products of the enzymatic reaction by fluorescence rather than by absorbance spectroscopy. This broad line of instrumentation allows the laboratory to perform several different types of assays using a single technology (EMIT).

IV. COMPARISON OF PERFORMANCE CHARACTERISTICS OF COMMERCIALLY AVAILABLE IMMUNOASSAY SYSTEMS FOR AMINOGLYCOSIDE AND VANCOMYCIN QUANTITATION

The preceding sections have reviewed the clinical and laboratory indications for monitoring serum levels of antimicrobial agents as well as the principles of the individual methodologies. In the subsequent sections I will provide a comparative analysis of the performance characteristics of the various assays. The data utilized in this analysis is that obtained from the many comparative studies reported in the literature and my own personal experience with the various methods. Because no one study has compared all of the available immunoassays the discussion must be of a general nature providing an overall opinion of the relative strengths and weakness of the various assays derived from data obtained from multiple individual studies.

A. Aminoglycoside Immunoassays

1. Comparison of General Features

The preceding sections have described the basic principles of six different types of immunoassays. These assays are available for the quantitation of the most commonly used aminoglycosides (gentamicin, tobramycin, amikacin) in serum and other body fluids. Table 1 presents a summary of the general features of each of these aminoglycoside assays. All utilize a specific polyclonal antibody as the basis for the assay. The method of detection is quite variable although four of the assays utilize fluorescence. Only the RIA utilizes radioactive material in any form. The antigen labels are indicated and range from none to the more complex β-galactosyl-umbelliferone conjugate. The detection range for all assays is similar. The range for gentamicin is given in Table 1 and generally is from 1 to 16 $\mu g/m\ell$.

Only two of the assays are heterogeneous requiring a separation step, while the remainder, including RNI, SLIFA, FPIA, and EMIT, are all homogeneous and require no separation of the product prior to measurement. As indicated previously, homogeneous assays offer significant advantages in speed, simplicity, and potential for automation.

Another important feature is the stability of the standard curve. This feature impacts on both the time required to perform an assay as well as on the cost of performing single (and multiple) assays. Thus, if a standard curve can be established independently of the individual test specimens on a daily or weekly basis, the ability to perform single or "STAT" analysis is greatly facilitated. The laboratory becomes much more flexible and can provide more individualized, and perhaps more clinically relevant, TDM service. The cost of establishing the standard curve can still be spread over multiple patient samples, as is done in batch testing, thus decreasing the cost of single sample testing. A total of three of the available aminoglycoside immunoassays, RIA, FIA, and SLIFA, require a standard curve with each run. All three methods either require a separation step or utilize a fluorescence based detection method. The immune-complex based assay (RNI) does not require that a standard curve be established but rather that a single calibrator be included with each run. The two assay methods with the longest standard curve stability are the homogeneous assays, FPIA and EMIT. The manufacturer recommends that the EMIT aminoglycoside standard curve be established daily; however, several investigators have documented that with proper storge of the reconstituted reagents (2 to 4°C) the standard curve for a given reagent lot is stable for 1 week or longer.[46-49] Likewise, there is considerable evidence that the standard curve for aminoglycoside determinations by FPIA is stable for 2 to 4 weeks.[37,41,43,49] The manufacturer currently recommends reestablishing the standard curve every 2 weeks for each aminoglycoside. It is strongly recommended that each laboratory determine the standard curve stability of each assay in their own unique setting in order to obtain optimal performance.

A significant factor in choosing an aminoglycoside assay is the instrumentation required. As indicated in Table 1 various types of instrumentation are necessary. Again, the individual needs and resources of the laboratory must be considered. However, the newer instrumentation is quite reliable and by utilizing homogeneous assays provides a degree of automation, speed, simplicity, and precision that is difficult or impossible to achieve with the less sophisticated systems and/or those employing heterogeneous assays. The adaptation of some of the homogeneous assay systems to highly automated enzyme analyzers has been shown to be extremely effective in improving the overall performance of the assay and reducing cost.[32,52-54]

2. Comparison of Specific Performance Characteristics

The information presented in Table 2 is a summary of data taken from several individual studies of aminoglycoside assay techniques over the time period 1980 to 1984. The aminoglycoside detected in most of these studies was gentamicin; however, there does not appear to be a significant difference in the performance of any given assay among the aminoglycosides. Each assay is ranked relative to the others for each of the specific performance categories described above. One must realize that in general all of the immunoassays are quite satisfactory for clinical use. The differences in performance indicated in this summary are a reflection of the overall trend observed in the literature and in our own laboratory studies and do not necessarily mean that one method is unsatisfactory or superior in a given situation. The relative performance of the immunoassays is discussed in the following sections.

a. Precision and Accuracy

As might be expected, the homogeneous automated systems such as RNI, FPIA, and EMIT provide superior precision compared to nonautomated homogeneous (SLIFA) and heterogeneous systems. Both within-run and between-run CVs for RNI, FPIA, and EMIT are less than 10% and usually in the 3 to 5% range.[28,37,38,46,48,49] The CVs for SLIFA, FIA, and RIA are comparable and range from 3 to 15%.[28,38,40] The adaptation of the SLIFA method to an automated centrifugal analyzer markedly improves the precision of the assay. Unfortunately, this type of instrumentation is not usually available to microbiology laboratory personnel.

The accuracy of all six methods is clinically acceptable. The percent recovery in serum samples spiked with a known amount of aminoglycoside is generally 100 ± 10%.[28,36-38,46,49] The correlation of RNI, FPIA, and EMIT with either theoretical concentrations in spiked samples or a reference bioassay or HPLC assay is in the range of r = 0.97 to 0.99 with minimal constant or proportional bias.[28,36-38,49] The remaining fluorescent methods (FIA, SLIFA) and the RIA all have comparable accuracy and tend to overestimate the actual concentration of the drug.[28,29,32,33,38]

b. Specificity

The specificity of all of the assay systems is excellent. Aside from the aminoglycosides there are little or no positive cross-reactions among a huge number of unrelated pharmacologic agents.[28,38,41] The RNI, FPIA, EMIT, and RIA systems are not significantly affected by hemolyzed, lipemic, or icteric sera.[28,36,38,41,55] The fluorescent assays (SLIFA and FIA) show a tendency to overestimate aminoglycoside levels in icteric sera[30-33,38] but this is usually adequately compensated for by substracting a blank reading from the final reading. Heparin has been shown to have a negative effect on the EMIT assays for tobramycin and gentamicin.[56] This effect is significant only at high levels of heparin (≥10 unit/mℓ for tobramycin and ≥100 units/mℓ for gentamicin) but should be kept in mind when dealing with specimens collected through a heparinized line or in a heparin-containing tube.[56] The FPIA and RIA are not effected by heparin.[56,57]

Another important drug-drug interaction affecting the aminoglycoside immunoassays is the interaction between β-lactam antibiotics and the aminoglycosides.[58-61] This interaction, which involves the penicillins (penicillin, ampicillin, carbenicillin, piperacillin, ticarcillin, azlocillin, mezlocillin) but not the cephalosporins, results in the biologic inactivation of both drugs.[58-61] Of the aminoglycosides tobramycin is the most susceptible and amikacin the least.[58-61] The inactivation takes place in vivo or in vitro and requires very high levels of the penicillin and/or prolonged incubation of the serum containing the two drugs at room temperature or above to produce a significant reduction in aminoglycoside concentration.[60] This interaction has been shown to affect the immunoassays for aminoglycosides as well as the bioassay.[58-61] To date this has been studied with the RIA, FPIA, and EMIT systems.[58-61] The inactivation detected by the RIA and FPIA is comparable and somewhat greater than that seen with the EMIT system.[58] The potential for this type of interaction should be considered with patients in renal failure (in vivo inactivation) or in situations in which improper handling or delays in transport result in prolonged incubation of serum samples at room temperature or greater (in vitro inactivation).[58,59,61]

c. Speed and Simplicity

In addition to showing increased precision, accuracy, and specificity, the three homogeneous automated methods, RNI, FPIA, and EMIT (with autocarousel), all are more simple to perform and generate results much faster than the heterogeneous and manual homogeneous (SLIFA) assays.[3,32,37,38] Once a standard curve has been established, both FPIA and EMIT are extremely rapid, taking approximately 1 to 2 min to perform each assay. Even the more manual versions of the EMIT system (without autocarousel) are more rapid than the heterogeneous assay systems. Likewise, the RNI with its preprogrammed standard curve requires only a single calibrator with each run and can produce a result in less than 10 min. All three systems are extremely simple to use, require minimal sample manipulation, and provide automatic data reduction. The RIA, FIA, and SLIFA (manual) all require extensive sample handling and dilution steps and relatively long incubation times.[28,32,38] They are not as simple or rapid as the other assays discussed above. The complexity of the procedures and the requirement of a standard curve with each run necessitates batch analyses which may further delay the reporting of individual test results. Again, the ability to adapt the homogeneous SLIFA to the automated centrifugal enzyme analyzers available in most clinical chemistry laboratories markedly simplifies and improves the performance of this test.[32]

d. Potential for Automation

It is apparent from the preceding sections that all of the homogeneous immunoassay methods can be easily automated. The instrumentation necessary to provide automated testing capabilities is available directly from the manufacturer for at least three of these assays (RNI, FPIA, and EMIT). The performance of the SLIFA and EMIT methods on centrifugal enzyme analyzers provides a high degree of automation with resultant improvement in speed, simplicity, and precision.[32,52-54] In addition, the reagent cost per test is decreased markedly.[52-54] In general there is little or no potential for automating the heterogeneous assays.

e. Application to Laboratory Workloads

The automated homogeneous assays are equally applicable to both large and small workloads as well as STAT testing.[28,32,37,38,48,49] Although smaller laboratories may find it difficult to justify the purchase of such instrumentation solely for the monitoring of aminoglycoside levels, it should be pointed out that these instruments generally can be used to perform a variety of other tests including monitoring of other pharmacologic agents and quantitation of a variety of serum proteins, hormones, and antibodies. All of the other systems are more applicable to larger workloads than smaller. STAT testing with these systems is both time consuming and expensive.

f. Cost Considerations

There are several cost considerations which must be weighed in selecting an immunoassay system for routine use in the laboratory. These are broken down into costs of equipment, reagents, and labor. Table 2 provides a ranking of each of the immunoassays as to relative cost in each of these three areas. In general, the more highly automated homogeneous methods require more expensive instrumentation and reagents than the other assay systems.[32,37,49] This is offset in part by a significant labor savings when automation is used. The effect of higher instrumentation and reagent costs on individual assay determinations is highly variable and dependent on several factors including workload, the type of testing (STAT vs. batch) performed, and the amount of hands-on technician time required to actually perform the assay.[28,37,38,49] Thus, although assays such as FPIA and EMIT may require more expensive equipment and reagents, the actual cost per patient specimen run may be quite comparable to that obtained with one of the less automated methods. This is particularly true when one considers the decreased number of standard curves and hands-on time required with the automated methods. Further adaptation of methods such as EMIT and SLIFA to rapid enzyme analyzers results in additional savings in reagents and time.[52-54]

Table 2
RELATIVE PERFORMANCE OF COMMERCIALLY AVAILABLE IMMUNOASSAY SYSTEMS FOR THE QUANTITATION OF AMINOGLYCOSIDES IN SERUM AND OTHER BODY FLUIDS[a,b]

Criterion	RIA	RNI	FIA	SLIFA[c]	FPIA	EMIT
Precision						
Within-run	++	++++	++	++	++++	++++
Between-run	++	++++	++	++	++++	++++
Accuracy	+++	++++	+++	+++	++++	++++
Specificity	++++	++++	++	++	++++	+++
Speed	+	+++	+	+	++++	+++
Simplicity	+	++++	+	+	++++	++++
Potential for automation	+	++++	+	++++	++++	++++
Applications						
Large workloads	+++	++++	+++	+++	++++	++++
Small workloads	++	+++	++	++	+++	+++
STAT analyses	+	++++	+	+	++++	++++
Cost considerations						
Equipment	++	+	+++	+++	+	+
Reagents	++	+++	+++	+++	+	++
Labor	++	++++	++	++	++++	++++

[a] Score: ++++ is better than +++ and so on. A score of + does not necessarily mean unsatisfactory performance.
[b] See Table 1 for abbreviations.
[c] Performed manually.

B. Vancomycin

1. Comparison of General Features

The assay techniques available to quantitate vancomycin in serum have included the bioassay, high-pressure liquid chromatography (HPLC), and radioimmunoassay (RIA; American Diagnostics, Newport Beach, Calif.).[39,45] Recently, two new immunoassays have also become available, one utilizing fluorescence polarization (FPIA; Abbott Laboratories) and the other a conventional fluorescence immunoassay (FIA; American Diagnostics).[39,44,45] The principles of each of these three immunoassays have been described previously in this chapter. The general features of each are summarized Table 3. The detection methods, antigen label, and instrumentation required for each assay are listed and the relative merits of each has been adequately discussed in the aminoglycoside section above (Section IV.A.1). Both the RIA and FIA methods for vancomycin determination require a separation step employing precipitation with a second antibody followed by centrifugation. The manipulations required of these heterogeneous assay systems is clearly a disadvantage in comparison to the homogeneous FPIA. The stability of the standard curve is another major difference between the FPIA and both RIA and FIA. The necessity to establish a new standard curve with each run is a major disadvantage of the latter two systems.[39] Finally, the detection range of the RIA is considerably less than either the FIA or FPIA.[39] This is due to the fact that the highest standard included with the RIA is 32 μg/mℓ. This means that all specimens with vancomycin concentrations of >32 μg/mℓ (approximately 15% of all specimens) must be diluted and rerun, a significant disadvantage.[39]

2. Comparison of Specific Performance Characteristics

There are relatively few studies comparing the performance characteristics of the

Table 3
FEATURES OF THREE VANCOMYCIN IMMUNOASSAYS

Test method[a]	Detection method	Antigen label	Separation step	Standard curve stability	Detection range (μg/mℓ)	Instrumentation
RIA (American Diagnostics)	Isotope	^{125}I	Yes, double antibody	Each run	1—32	Gamma counter
FIA (American Diagnostics)	Fluorescence	Fluorescein	Yes, double antibody	Each run	1—128	Fluorometer
FPIA (Abbott)	Fluorescence	Fluorescein	No	2—4 weeks	1—100	Fluorescence polarization analyzer

[a] See Table 1 for abbreviations.

Table 4
RELATIVE PERFORMANCE OF COMMERCIALLY AVAILABLE IMMUNOASSAY SYSTEMS FOR THE QUANTITATION OF VANCOMYCIN IN SERUM AND OTHER BODY FLUIDS[a,b]

Criterion	RIA	FIA	FPIA
Precision			
Within-run	+++	+	++++
Between-run	+++	+	++++
Accuracy	++++	++++	++++
Specificity	++++	+++	++++
Speed	+	+	++++
Simplicity	+	+	++++
Potential for automation	+	+	++++
Applications			
Large workloads	+++	+++	++++
Small workloads	++	++	+++
STAT analyses	+	+	++++
Cost considerations			
Equipment	++	++	+
Reagents	++	++	+
Labor	++	++	++++

[a] Score: see Table 2.
[b] Abbreviations: see Table 1.

various assays for vancomycin quantitation[39,42,44,45,62] and only one which provides a comparison of all three commercially available immunoassays.[39] Table 4 provides an overview of the relative performance of each of these assays. As expected from the previous discussion of aminoglycoside assays, the FPIA for vancomycin is clearly superior to the FIA and RIA in almost every category. The precision of the FPIA is excellent over the therapeutic range (7 to 75 μg/mℓ). Within-run CVs are ≤3% for FPIA compared with 4.7 to 10.6% for RIA and 8.9 to 14.5% for FIA.[39,44] Between-run CVs range from 2.8 to 8.1% for FPIA vs. 7.3 to 11.0% for RIA and 12.2 to 16.2%

for FIA.[39,44] All three methods are highly accurate and have excellent specificity.[39,42,44,45,62] As with the aminoglycoside assays, the vancomycin FPIA is clearly superior to FIA and RIA in speed, simplicity, potential for automation, and application to a wide range of workloads. The cost considerations are similar to those discussed with the aminoglycoside assays with saving of labor being the major advantage of FPIA.

V. SUMMARY AND CONCLUSIONS

The availability of new immunoassay methods has led to a revolution in therapeutic drug monitoring. This has resulted in a marked increase in the number of serum aminoglycoside and vancomycin levels performed in both microbiology and clinical chemistry laboratories. The newer homogeneous immunochemical methods for quantitation of aminoglycosides (FPIA, EMIT, RNI, and SLIFA) and vancomycin (FPIA) meet the analytical requirements of specificity, precision, and accuracy necessary for most clinical applications. The ability to automate these methods has further improved their precision and ease of performance. Overall it appears that two assay systems, EMIT (Syva) and FPIA (Abbott TDX®; Abbott Laboratories), offer superior precision, speed, simplicity, and flexibility compared to the other immunoassay systems for the determination of aminoglycoside (EMIT and FPIA) and vancomycin (FPIA only) levels in clinical specimens. The technique chosen for each laboratory will depend upon numerous factors including the availability of personnel and equipment, the time required to perform the assay, the speed with which the clinician needs the results, and the cost and medical benefit to the patient. Regardless of the assay technique used in the laboratory, efforts should be made to ensure proper utilization of this valuable clinical information.

ACKNOWLEDGMENTS

The author would like to thank Mr. Tom Persoon and Dr. George Johnson of the University of Iowa and Mr. George Granich, Dr. Patrick Murray, and Dr. Donald Krogstad of Barnes Hospital and Washington University in St. Louis for their help and suggestions. In addition, the help and encouragement of Dr. Robert Kreisler of International Research Consultants was greatly appreciated.

REFERENCES

1. Koch-Weser, J., Serum drug concentrations in clinical perspective, in *Therapeutic Drug Monitoring,* Richens, A. and Marks, V., Eds., Churchill Livingstone, Edinburgh, 1981, 1.
2. Marks, V., Immunoassays for drugs, in *Therapeutic Drug Monitoring,* Richens, A. and Marks, V., Eds., Churchill Livingstone, Edinburgh, 1981, 155.
3. Munro, A. J., Landon, J., and Shaw, E. J., The basis of immunoassays for antibiotics, *J. Antimicrob. Chemother.,* 9, 423, 1982.
4. Matzke, G. R., Lucaroth, P. L., and Shapiro, H. S., Controlled comparison of gentamicin and tobramycin nephrotoxicity, *Am. J. Nephro.,* 3, 11, 1983.
5. Jackson, G. G. and Arcieri, G., Ototoxicity of gentamicin in man: a survey and controlled analysis of clinical experience in the United States, *J. Infect. Dis.,* 124(Suppl.), S130, 1971.
6. Zaske, D. E., Cipolle, R. J., Solena, L. D., and Strate, R. G., Rapid individualization of gentamicin dosage requirements in 66 burn patients, *Burns,* 7, 215, 1981.
7. Zaske, D. E., Cipolle, R. J., and Strate, R. G., Gentamicin dosage requirements: wide interpatient variations in 242 surgery patients with normal renal function, *Surgery,* 87, 164, 1980.

8. Lesar, T. S., Rotschafer, J. C., Strand, L. M., Solem, L. D., and Zaske, D. E., Gentamicin dosing errors with four commonly used nomograms, *JAMA*, 248, 1190, 1982.
9. Zaske, D. E., Bootman, J. L., Solem, L. B., and Strate, R. G., Increased burn patient survival with individualized dosages of gentamicin, *Surgery*, 91, 142, 1982.
10. Cipolle, R. J., Seifert, R. D., Zaske, D. E., and Strate, R. G., Hospital acquired gram-negative pneumonias: response rate and dosage requirements with individualized tobramycin therapy, *Ther. Drug Monit.*, 2, 273, 1980.
11. Moore, R. D., Smith, C. R., and Lietman, P. S., The association of aminoglycoside plasma levels with mortality in patients with gram-negative bacteremia, *J. Infect. Dis.*, 149, 443, 1984.
12. Farber, B. F. and Moellering, R. C., Jr., Retrospective study of the toxicity of preparations of vancomycin from 1974 to 1981, *Antimicrob. Agents Chemother.*, 23, 138, 1981.
13. Moellering, R. C., Jr., Krogstad, D. J., and Greenblatt, D. J., Vancomycin therapy in patients with impaired renal function: a nomogram for dosage, *Ann. Intern. Med.*, 94, 343, 1981.
14. Rotschafer, J. C., Crossley, K., Zaske, D. E., Mead, K., Sawchuk, R. J., and Solem, L. D., Pharmacokinetics of vancomycin: observations in 28 patients and dosage recommendations, *Antimicrob. Agents Chemother.*, 22, 391, 1982.
15. Blevins, R. D., Halstenson, C. E., Salem, N. G., and Matzke, G. R., Pharmacokinetics of vancomycin in patients undergoing continuous ambulatory peritoneal dialysis, *Antimicrob. Agents Chemother.*, 25, 603, 1984.
16. Blouin, R. A., Bauer, L. A., Miller, D. D., Record, K. E., and Griffin, W. O., Jr., Vancomycin pharmacokinetics in normal and morbidly obese subjects, *Antimicrob. Agents Chemother.*, 21, 575, 1982.
17. Anderson, A. C., Hodges, G. R., and Barnes, W. G., Determination of serum gentamicin sulfate levels, *Arch. Intern. Med.*, 136, 785, 1976.
18. Flynn, T. W., Pevonka, M. P., Yost, R. L., Weber, C. E., and Stewart, R. B., Use of serum gentamicin levels in hospitalized patients, *Am. J. Hosp. Pharm.*, 35, 806, 1978.
19. Bussey, H. I. and Hoffman, E. W., A prospective evaluation of therapeutic drug monitoring, *Ther. Drug Monit.*, 5, 245, 1983.
20. Elenblas, R. M., Payne, V. W., and Bauman, J. L., Influence of clinical pharmacist consultations on the use of drug blood level tests, *Am. J. Hosp. Pharm.*, 37, 61, 1980.
21. Lampasona, V. and Crass, R. E., Patient selection for serum gentamicin levels, *Ther. Drug Monit.*, 5, 255, 1983.
22. Matzke, G. R., Halstenson, C. E., and Keane, W. F., Hemodialysis elimination rates and clearance of gentamicin and tobramycin, *Antimicrob. Agents Chemother.*, 25, 128, 1984.
23. Sampliner, R., Perrier, D., and Finley, P., Influence of ascites on tobramycin pharmacokinetics, *J. Clin. Pharmacol.*, 24, 43, 1984.
24. Sketris, I., Lesor, T. S., Zaske, D. E., and Cipolle, R. J., Effect of obesity on gentamicin pharmacokinetics, *J. Clin. Pharmacol.*, 21, 288, 1981.
25. Schentag, J. J., Cumbo, T. J., Jusko, W. J., and Plaut, M. E., Gentamicin tissue accumulation and nephrotoxic reactions, *JAMA*, 240, 2067, 1978.
26. Gallagher, K. L. and Jones, J. K., Furosemide-induced ototoxicity, *Ann. Intern. Med.*, 91, 744, 1979.
27. Butler, W. T., Bennett, J. E., Alling, D. W., Wertlake, P. T., Utz, J. P., and Hill, G. J., Nephrotoxicity of amphotericin B, early and late effects in 81 patients, *Ann. Intern. Med.*, 61, 175, 1964.
28. Ratcliff, R. M., Mirelli, C., Moran, E., O'Leary, D., and White, R., Comparison of five methods for the assay of serum gentamicin, *Antimicrob. Agents Chemother.*, 19, 508, 1981.
29. Stobberingh, E. E., Houben, A. W., and Van Boven, C. O. A., Comparison of different tobramycin assays, *J. Clin. Microbiol.*, 15, 795, 1982.
30. Hicks, J. M., Fluorescence immunoassay, *Human Pathol.*, 15, 112, 1984.
31. Landon, J. and Kamel, R. S., Immunoassays employing reactants labeled with a fluorephore, in *Immunoassays for the 80s*, Voller, A., Bartlett, A., and Bidwell, D., Eds., University Park Press, Baltimore, 1981, 91.
32. Boguslaski, R. C. and Burd, J. F., Analytical methods for therapeutic drug monitoring, *Am. J. Med. Technol.*, 49, 551, 1983.
33. Thompson, S. G., Homogeneous substrate-labeled fluorescent immunoassays for determining levels of low molecular weight substances in human serum, in *Nonisotopic Alternatives to Radioimmunoassay: Principles and Applications*, Kaplan, L. A. and Pesce, A. J., Eds., Marcel Dekker, Inc., New York, 1981, 107.
34. Spencer, R. D., Applications of fluorescent polarization in clinical assays, in *Nonisotopic Alternatives to Radioimmunoassay: Principles and Applications*, Kaplan, L. A. and Pesce, A. J., Eds., Marcel Dekker, Inc., New York, 1981, 143.
35. Gauldie, J., Principles and clinical applications of nephelometry, in *Nonisotopic Alternatives to Radioimmunoassay: Principles and Applications*, Kaplan, L. A. and Pesce, A. J., Eds., Marcel Dekker, Inc., New York, 1981, 285.

36. Deaton, C., Derry, C., and Decker, M., A nephelometric inhibition immunoassay system for the therapeutic monitoring of anticonvulsants, aminoglycosides, and theophylline, *Clin. Chem.*, 28, 1611, 1982.
37. Amsterdam, D., Kalinka, C., and Bartholomew, W. R., Comparative immunoassays for quantitation of tobramycin serum levels, *Antimicrob. Newsl.*, 1, 56, 1984.
38. Selepack, S. T., Witebsky, F. G., Robertson, E. A., and MacLowry, J. D., Evaluation of five gentamicin assay procedures for clinical microbiology laboratories, *J. Clin. Microbiol.*, 13, 742, 1981.
39. Pfaller, M. A., Krogstad, D. J., Granich, G. G., and Murray, P. R., Laboratory evaluation of five assay methods for vancomycin: bioassay, high-pressure liquid chromatography, fluorescence polarization immunoassay, radioimmunoassay, and fluorescence immunoassay, *J. Clin. Microbiol.*, 20, 311, 1984.
40. Burd, J. F., Wong, R. C., Feeney, J. D., Carrico, R. J., and Boguslaski, R. C., Homogeneous reactant-labeled fluorescent immunoassay for therapeutic drugs exemplified by gentamicin determination in human serum, *Clin. Chem.*, 23, 1402, 1977.
41. Jolley, M. E., Stroupe, S. D., Wang, C. J., Panas, H. N., Keegan, C. L., Schmidt, R. L., and Schwenzer, K. S., Fluorescence polarization immunoassay. I. Monitoring aminoglycoside antibiotics in serum and plasma, *Clin. Chem.*, 27, 1190, 1981.
42. Filburn, B. H., Shull, V. H., Tempera, Y. M., and Dick, J. D., Evaluation of an automated fluorescence polarization immunoassay for vancomycin, *Antimicrob. Chemother.*, 24, 216, 1983.
43. Oeltgen, P. R., Shank, W. A., Blouin, R. A., and Clark, T., Clinical evaluation of the Abbott TDX fluorescence polarization immunoassay analyzer, *Ther. Drug Monit.*, 6, 360, 1984.
44. Schwenzer, K. S., Wang, C. J., and Anholt, J. P., Automated fluorescence polarization immunoassay for monitoring vancomycin, *Ther. Drug Monit.*, 5, 341, 1983.
45. Ristuccia, P. A., Ristuccia, A. M., Bidanset, J. H., and Cunha, B. A., Comparison of bioassay, high-performance liquid chromatography, and fluorescence polarization immunoassay for quantitative determination of vancomycin in serum, *Ther. Drug Monit.*, 6, 238, 1984.
46. Delaney, C. J., Opheim, K. E., Smith, A. L., and Plorde, J. J., Performance characteristics of bioassay, radioenzymatic assay, homogeneous enzyme immunoassay, and high-performance liquid chromatographic determination of serum gentamicin, *Antimicrob. Agents Chemother.*, 21, 19, 1982.
47. Buchanan, A. G., Witwicki, E., and Albritton, W. L., Serum aminoglycoside monitoring by enzyme immunoassay, biological, and fluorescence immunoassay procedure, *Am. J. Med. Technol.*, 49, 437, 1983.
48. Pohlod, D. J., Saravolatz, L. D., and Somerville, M. M., Comparison of enzyme-multiplied immunoassay technique with fluorescence polarization immunoassay for determination of gentamicin and tobramycin levels in serum, *J. Clin. Microbiol.*, 20, 866, 1984.
49. Ngui-Yen, J. H., Doyle, P. W., and Smith, J. A., Comparative analysis of two rapid, automated methods for determining aminoglycoside levels, *J. Clin. Microbiol.*, 20, 962, 1984.
50. Rubenstein, K. E., Schneider, R. S., and Ullman, E. F., "Homogeneous" enzyme immunoassay. A new immunochemical technique, *Biochem. Biophys. Res. Commun.*, 47, 846, 1972.
51. Rowley, G. L., Rubenstein, K. E., Huisjen, J., and Ullman, E. F., Mechanism by which antibodies inhibit hapten-malate dehydrogenase conjugates, *J. Biol. Chem.*, 250, 3759, 1975.
52. Hamlin, C. R. and Sullivan, P. A., Adaptation of EMIT reagents to the Cobas Bio centrifugal analyzer, *Clin. Chem.*, 30, 314, 1984.
53. Shaw, W. and McHan, J., Adaptation of EMIT procedures for maximum cost effectiveness to two different centrifugal analzyer systems, *Ther. Drug Monit.*, 3, 185, 1981.
54. Iosefsohn, M., Boeckx, R. L., and Hicks, J. M., Adaptation of the EMIT gentamicin and tobramycin procedure to the IL microcentrifugal analyzer, *Ther. Drug Monit.*, 3, 365, 1981.
55. Stewart, C. F., Melton, E. T., Karas, J., and Evans, W. E., A randomized-blinded evaluation of analytical error associated with the enzyme immunoassay (EMIT) of gentamicin, *Ther. Drug Monit.*, 4, 409, 1982.
56. Krogstad, D. J., Granich, G. G., Murray, P. R., Pfaller, M. A., and Valdes, R., Heparin interferes with the radio-enzymatic and homogeneous enzyme immunoassays for aminoglycosides, *Clin. Chem.*, 28, 1517, 1982.
57. Jackson, D. S. and Bertino, J. S., Jr., Effects of heparin sodium on gentamicin assay measured by fluorescence polarization immunoassay method, *Ther. Drug Monit.*, 6, 319, 1984.
58. Pfaller, M. A., Granich, G. G., Valdes, R., and Murray, P. R., Comparative study of the ability of four aminoglycoside assay techniques to detect the inactivation of aminoglycosides by β-lactam antibiotics, *Diagn. Microbiol. Infect. Dis.*, 2, 93, 1984.
59. Henderson, J. L., Polk, R. E., and Kline, B. J., In-vitro inactivation of gentamicin, tobramycin, and netilmicin by carbenicillin, azlocillin, or mezlocillin, *Am. J. Hosp. Pharm.*, 38, 1167, 1981.
60. Pickering, L. K. and Rutherford, I., Effect of concentration and time upon inactivation of tobramycin, gentamicin, netilmicin and amikacin by azlocillin, carbenicillin, mecillinam, mezlocillin and piperacillin, *J. Pharmacol. Exp. Ther.*, 217, 345, 1981.

61. Animal model distinguishing in vitro from in vivo carbenicillin-aminoglycoside interactions, *Antimicrob. Agents Chemother.*, 18, 604, 1980.
62. Ackerman, B. H., Berg, H. G., Strate, R. G., and Rotschafer, J. C., Comparison of radioimmunoassay and fluorescent polarization immunoassay for quantitative determination of vancomycin concentrations in serum, *J. Clin. Microbiol.*, 18, 994, 1983.

Chapter 11

INSTRUMENT-BASED SERODIAGNOSTIC METHODS

James E. Johnson

TABLE OF CONTENTS

I. INTRODUCTION

Serodiagnosis for many infectious diseases will continue to be of value in the future. Serologic testing may be the only or the most rapid way to diagnose infectious diseases when slow-growing or nonculturable microorganisms including bacteria, fungi, and viruses are the cause of an illness or when patients have been partially treated with antibiotics. Serologic testing can also document infection vs. colonization and determine past from current infection. In addition, serologic testing may be used to rapidly detect infectious antigens directly from various body fluids. The development and increased use of monoclonal antibodies will increase both sensitivity and specificity of serodiagnostic procedures while improvements in automation will further increase test sensitivity and reproducibility.

Major efforts have been made in development of automated instruments for the clinical laboratory, especially chemistry and hematology analyses. However, until recently, few instruments were available for use in microbiology and serology. While there have been many reasons for a lack of automation in serology, utilization of mechanized and/or automated equipment for the serodiagnosis of infectious diseases should increase objectivity and reproducibility, decrease technologist time, and shorten total turnaround time in serologic testing. Instrumentation has been developed for use in miniaturized serologic testing systems, i.e., microdilution tests. Various microtiter pipettors, dilutors, washing systems, and readers are now available for use with complement fixation (CF) and hemagglutination procedures. Equipment originally developed for clinical chemistry can be utilized in testing of sera by radioimmunoassay (RIA) and enzyme immunoassay (EIA) techniques. Fluorescent immunoassay (FIA) procedures have also been partially automated through the use of a solid phase fluoroimmunoassay system.

The purpose of this chapter will be to discuss a number of the commercially available systems based on type of antibody detection: RIA, FIA, or EIA.

II. TEST METHODOLOGY

A. Radioimmunoassay (RIA)

The best-known application of RIA for the serodiagnosis of infectious diseases is testing for hepatitis B surface antigen (HB_sAg) and antibody (anti-HB_sAg). These tests are based primarily on studies with a radioimmunoprecipitin technique described in 1970.[1] Currently available commercially produced kits utilize a solid phase system made possible by the discovery of Catt and Tregear that proteins strongly adhere to plastic and glass surfaces and cannot be readily washed off.[2] In testing for the presence of anti-HB_sAg, (Figure 1) the solid phase (disc, tube, microtiter well, or bead) is coated with HB_sAg. After the patient's serum is added, if antibody to HB_sAg is present in the patient's serum, it will bind to the HB_sAg attached to the solid phase. After washing, ^{125}I-labeled HB_sAg is added which will in turn bind to the antibody from the patient's serum, thus forming an antigen-antibody "sandwich". Following appropriate incubation and washing, positive results are indicated by the use of a gamma scintillation counter to determine the level of bound isotope. In like manner, when detecting antigen in serum, antibody is attached to the solid phase and an ^{125}I-labeled antibody is used as the indicator of a positive reaction. While the procedure can be semiautomated through the use of pipettors, dilutors, and washing devices, the requirement for a scintillation counter makes this test a relatively expensive procedure. Other disadvantages include the necessity for handling and disposal of radioactive material, as well as the short shelf-life of isotopic reagents.

1. SP-A + B ⟶ SP-A-B

2. SP-A-B + A* ⟶ SP-A-B-A*

FIGURE 1. Basic heterogeneous radioimmunoassay antibody detection scheme. SP = solid phase; A = antigen; B = antibody; A* = radiolabeled antigen.

Although RIA procedures have been developed for the serodiagnosis of many bacterial,[3] viral,[4] and parasitic diseases,[5] the majority of those tests have not been commercially marketed and, thus, have not gained wide acceptance and use. Recently an RIA for the detection of a tuberculoprotein has been developed which may serve as a model for the development of protein assays to speed up the diagnosis of infectious diseases caused by slow-growing microorganisms such as *M. tuberculosis*.[6,7]

B. Fluoroimmunoassay (FIA)

Immunofluorescence, both direct and indirect methodologies, have been and still are the methods of choice for the serodiagnosis of many infectious diseases, e.g., rabies, legionellosis, Epstein-Barr virus associated infections, rickettsial diseases, etc. The indirect immunofluorescent test (IIF) is the version of the test used for antibody detection. In the IIF procedure, an antigen is fixed to a microscope slide, a number of dilutions of patient's serum are added, and subsequently, an antihuman immunoglobulin tagged with fluorescein isothiocyanate (FITC) is added to the wells. Upon appropriate incubation and washing steps, the slide is ready for microscopic examination using a UV microscope. Although IIF has been readily accepted for use in the serology laboratory, it is a labor intensive procedure with a high degree of subjectivity involved in interpretation of end-points.

1. FIAX

The standard IIF test has been mechanized by International Diagnostics Technology (IDT, Whittaker M. A. Bioproducts, Walkersville, Md.). Solid-phase fluoroimmunoassays were originally developed for the determination of antinuclear antibody and immunoglobulin levels using the FIAX 100 digital fluorometer in conjunction with the FIAX 110 microcomputer. FIAX assays are similar to the IIF procedure described above. However, the microscope slide is replaced by a unique STIQ™ sampler on which an antigen has been absorbed. The STIQ™ is immersed in a single dilution of patient's serum, incubated, washed and then reacted with FITC-labeled anti-human immunoglobulin. After incubation and a final wash step, fluorescence is determined by inserting the STIQ™ into the central port of the fluorometer. An automatic readout of the degree of fluorescence is determined and a calculation is performed by comparison to a standard curve to yield a quantitative result. The major advantages of the FIAX system are the high degree of objectivity in the determination of a positive result, the use of a single serum dilution for determination of a precise end-point rather than a titer range, and minimal hands-on time (less than 1 min) per sample. Currently, FIAX tests available for serodiagnosis of infectious diseases include rubella IgG and IgM antibodies, herpes simplex IgG antibodies, cytomegalovirus IgG antibodies, and toxoplasma IgG antibodies. Tests for IgM antibodies for both cytomegalovirus and *Toxoplasma gondii* are being developed. Evaluations of the above products indicate that the FIAX system is sensitive, specific, rapid, reproducible, and economical.[8-13] Taylor and Perez developed a qualitative test for the serodiagnosis of amoebiasis using the FIAX which has subsequently been modified to give quantitative results.[14,15] Walls and Wilson compared the quantitative FIAX test for amoebiasis with an indirect hemagglutination test (IHA).[15] Their results indicated the FIAX method to be more sensitive than IHA, based upon the ability to detect chronic cases, as well as being more specific than IHA. The FIAX test demonstrated only 2% false-positives at a dilution of 1:128 compared to 18% false-positive reactions at the same dilution for the IHA test. Stevens has reported on the development of tests for the diagnosis of syphilis and Legionnaires' disease using the FIAX.[16] Although these latter tests are not yet commercially available, it would appear that a large number of microbial antigens could be absorbed to the STIQ™ sampler and thus allow development of still other diagnostic tests.

IDT is now introducing an updated version of the FIAX called the FIAX 400 system. It consists of a fluorometer with a built in microprocessor, a dilutor, and a shaker. Both the fluorometer and dilutor are programmed by magnetic cards that are provided with each reagent kit. The system also has the capability of interfacing with an external computer for automated data handling.

2. Fluorescence Polarization (FP)

Fluorescence polarization immunoassays have been developed primarily for therapeutic drug monitoring, including antibodies (see Chapter 10). The technique used is a competitive binding immunoassay in which fluorescein-labeled drug absorbs polarized light and the degree of polarization of the subsequent emitted light is measured. Unlabeled drug in the patient's serum competes with the labeled drug for binding sites on the antibodies. The rate of rotation of bound drug is slower than unbound, increasing the retained polarity of the light. The change in polarization is measured and used to calculate the amount of drug present. Initial problems with this technique were the requirement for a relatively long lifetime of the fluorescent probe, background light interference, and nonlinear response in relation to drug concentration.[17] The TDX® system (Abbott Laboratories, Irving, Tex.), uses microprocessor technology in conjunction with a series of drug standards to calculate a curve fit algorithm which overcomes the aforementioned problems. To date, tests for determining the presence of microbial antibodies or antigens have not been developed for commercial use. However, at least in theory, FP methodology is applicable for measurement of virtually all antigens and antibodies.[18,19]

3. Time-Resolved Fluoroimmunoassay (TR-FIA)

One of the major problems with conventional immunofluorescence testing is the high background fluorescence caused by serum and other interfering materials. Time-resolved fluorescence is a method in which the effect of background fluorescence inherent in FIA procedures is negated. The TR-FIA principle, as described by Soini and Hemmila,[17] involves measuring the fluorescence of the probe after sufficient time has elapsed from the moment of excitation so that short-lived background fluorescence (less than 10 nsec) has decayed. This requires that the fluorescent probe have a long decay time, generally greater than 100 nsec (Figure 2). Immunoglobulins labeled with europium chelates appear to give sensitivities at least equal to those seen in RIA procedures for detection of viral antigens.[20] These chelates can be easily measured with a xenon discharge lamp or through the use of a nitrogen laser.[17] In 1982, it was reported that rubella antibodies could be detected with a sensitivity and specificity comparable to RIA and EIA.[21] In this study, polystyrene beads coated with rubella antigen were used as the solid phase and a europium chelate was used as the fluorescent label. A fluorometer (developed by LKB Wallac, Turku, Finland) with a xenon flash lamp was used to measure photon emission over a 1 sec counting time with excitation pulses given

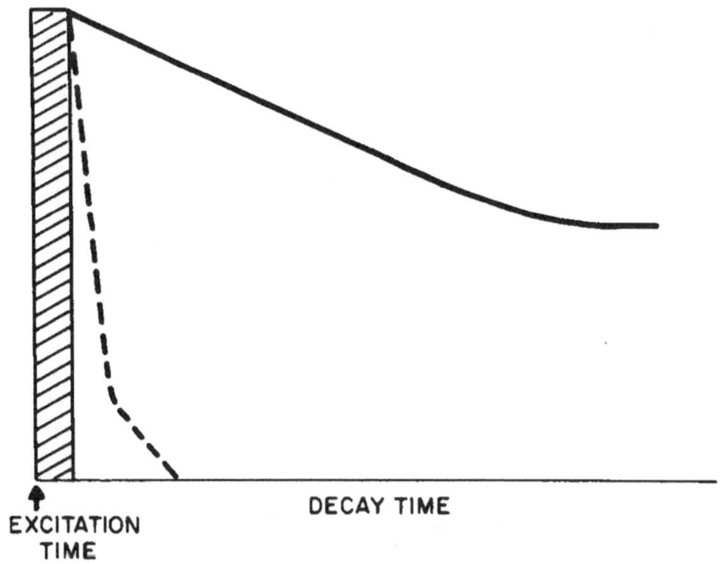

FIGURE 2. Graphical representation of time-resolved fluoroimmunoassay theory. —— = Probe fluorescence; ----- = background fluorescence.

at 100 μsec intervals. An arbitrary TR-FIA unit was developed by making a standard curve. Overall agreement with a hemagglutination test and a commercially available RIA procedure was 96.7%. The authors concluded that TR-FIA could be an acceptable alternative to RIA for the detection of microbial antigens and antibodies.

4. MicroFLUOR™

The microFLUOR™ system (Dynatech Laboratories, Inc., Alexandria, Va.) features 96-well low fluorescence microtiter plates which can be objectively read in the microFLUOR™ reader. The system, as currently designed, uses methylumbelliferone as the fluorescent label and reports results as relative fluorescent units. The manufacturer's literature states that the microFLUOR™ system provides sensitivity and precision equal to that of RIA testing with none of the storage and disposal problems associated with RIA. To date, the system has been used solely for IgE allergy testing, although it has the potential to be used for infectious disease serodiagnosis. For this to occur, the system must develop more versatility so that multiple wavelengths and other fluorescent compounds can be used.[22] Currently, a single wavelength of 450 nm is available.

5. Homogeneous Fluoroimmunoassay (HFA)

Most RIA and ELISA procedures are heterogeneous in nature. They require separation of bound and free antigen or antibody by washing prior to measuring the products of the reaction. A homogeneous assay is one in which the product can be measured in the presence of the reactants. The Ames TDA® system (Ames Laboratories, a subsidiary of Miles Laboratories, Inc., Elkhart, Ind.) was designed for therapeutic drug assays, primarily aminoglycoside antibiotics (see Chapter 10). Subsequent to the development of the technique of homogeneous reactant-labeled fluorescent immunoassay,[23] a fluorometer has been introduced for use with the TDA® system. The principle of the assay is competitive protein binding between the patient's drug and a known amount of drug labeled with umbelliferyl-βD-galactoside (Figure 3). The labeled substrate/drug combination is nonfluorescent. Both the drug present in the patient's

1. $\text{GUD} \xrightarrow{E} \text{UD}^*$
2. $\text{GUD} + \text{B} \longrightarrow \text{GUD-B} + \text{D-B} + \text{UD}^*$
3. $\text{GUD-B} \xrightarrow{E} \text{NR}$
4. $\text{GUD} + \text{B} + \text{D} \xrightarrow{E} \text{GUD-B} + \text{D-B} \_ \text{UD}^*$

FIGURE 3. Schematic representation of reactions involved in a homogeneous fluoroimmunoassay. GUD = umbelliferyl-β-D-galactoside; E = the enzyme, β-galactosidase; UD* = fluorescent unbound drug; B = antibody to drug; NR = no reaction; D-B = unlabeled drug-antibody complex.

1. $\text{SP-A} + \text{B} \longrightarrow \text{SP-A-B}$
2. $\text{SP-A-B} + \text{B}^* \longrightarrow \text{SP-A-B-B}^*$
3. $\text{SP-A-B-B}^* + \text{S} \longrightarrow \text{C}$

FIGURE 4. Heterogeneous sandwich enzyme immunoassay for the detection of antibody. SP = solid phase; A = antigen; B = antibody; B* = enzyme labeled antiimmunoglobulin; S = substrate; C = colored complex.

serum and the enzyme substrate labeled drug are mixed with specific antibody to the drug and, therefore, compete with each other for binding sites on the antibody. Depending upon the amount of drug in the patient's serum, a certain amount of substrate-labeled drug will remain unbound. When the enzyme, β-galactosidase is added to the mixture, the unbound labeled drug is broken down, forming a fluorescent compound. The resulting fluorescence is proportional to the amount of the patient's drug in the specimen. The use of this system for the measurement of aminoglycosides in human serum indicates the TDA® system is relative simple, uses stable reagents, possesses sensitivity and specificity similar to RIA, and may be applicable to a large number of biologically active molecules.[23] It appears that the HFA has the potential to be used for detection of antibodies found in response to various infectious diseases.

C. Enzyme Immunoassay (EIA)

Enzyme immunoassay tests have proven to be reliable alternatives to RIA with the primary advantages being no requirement for radioactive materials and the longer shelf life of EIA reagents. EIA is a general term for a number of different assays utilizing enzymes as part of the detection system. Homogeneous enzyme immunoassays may specifically be called enzyme-multiplied immunoassay techniques (EMIT). The heterogeneous enzyme immunoassays which utilize a solid phase support system are collectively termed enzyme-linked immunosorbent assays (ELISA). The most common ELISA methods used for serodiagnosis of infectious diseases are "sandwich" antigen assays and "sandwich" antibody assays. Figure 4 demonstrates the steps involved in a sandwich antibody assay. The use of an enzyme labeled anti-immunoglobulin rather than an enzyme-labeled antigen seems to increase the sensitivity of the assay.[24] A number of review articles have described the numerous EIA procedures which are available.[25-29] The heterogeneous solid phase assays are similar in sensitivity and specificity to RIA methods and are performed in the same manner with the requirement for one

Table 1
SOME COMMERCIALLY AVAILABLE EIA KITS AND THEIR SUPPLIERS

Manufacturer	Product[a]
Abbott	Antibody detection for anti HB_sAg, e, total core, and core IgM, hepatitis A IgM, rubella IgG and IgM, and CMV; antigen detection for HB_sAg, rotavirus, chlamydia and *Neisseria gonorrhea*
Behring	Antibody detection for rubella IgG and IgM, CMV IgG and IgM, HSV IgG and IgM; antigen detection for rotavirus (antigen detection for HB_sAg and antibody detection for anti HB_sAg, measles, mumps, and varicella-zoster available only in Europe)
Cordis	Antibody detection for toxoplasma IgG and IgM, rubella IgG and IgM, CMV, HSV, amoebiasis, syphilis, chlamydia; antigen detection for HB_sAg, HSV, and *N. gonorrhea*
Whittaker M. A. Bioproducts	Antibody detection for HSV, CMV IgG and IgM, rubella IgG and IgM, mumps, measles, varicella-zoster, mycoplasma, toxoplasma, chlamydia.
Lab Systems, Inc.	Antibody detection for toxoplasma IgG and IgM, CMV IgG and IgM, rubella, pertussis, tetanus toxoid.
Litton Bionetics	Antibody detection for toxoplasma IgG and IgM, rubella, CMV, HSV, syphilis, amoebiasis, and varicella-zoster; antigen detection for rotavirus
Ortho Diagnostics	Antibody detection for rubella, CMV, toxoplasma, HSV-1, HSV-2; antigen detection for HSV, RSV.

[a] Only those products related to serodiagnosis of infectious diseases are listed.

additional step, the adding of the enzyme to the test system. A recent study,[30] using aliquots of antihuman IgG labeled with alkaline phosphatase or ^{125}I, allowed the direct comparison of ELISA and RIA. Both direct and indirect assays were performed with comparable results noted with both systems. Since ELISA methodology is comparable to RIA, the same types of solid-phase supports may be used, including plastic beads, tubes, cuvettes, microtiter trays, etc. The most common enzyme/substrate systems currently used for serodiagnosis are alkaline phosphatase/*p*-nitrophenyl phosphate and horseradish peroxidase/ϕ-phenylene-diamine. Most of the limiting factors involved in EIA have been successfully dealt with in the commercial development of both antigen and antibody detection systems. Although many microbial antigens and antibodies have been evaluated in a research setting, Table 1 lists those ELISA procedures which are commercially marketed and their suppliers.

The following section will discuss instrumentation developed by the companies listed in Table 1. It should be mentioned that other systems are available which are either semiautomated or totally automated and that many of these are currently used in clinical chemistry laboratories for EIA testing. Many of these have yet to be adapted for kits capable of detecting antigen/antibody reactions for serodiagnosis. An excellent pictorial review of many semi- and totally automated systems used in viral diagnosis was published by Sever.[31]

1. Quantum II

The Quantum II (Abbott Laboratories, North Chicago, Ill.) is the updated version of the Quantum I, described by Sever.[31] It is a dual wavelength analyzer designed to read single tube ELISA tests at a set wavelength of 492 nm. The system can read one tube every second. The Quantum II data processor consists of a microcomputer and preprogrammed interchangeable memory modules, which store the calculation steps used for each of the assays. These programs can also be modified by the operator. Major advantages of this system are that the Quantum II is provided to customers as part of the reagent purchase, data reduction for all Abbott EIA tests is provided thereby saving technician time, and software and hardware upgrades are provided by the manufacturer.

The Abbott EIA test procedures utilize a microtiter tray designed to hold 20 polystyrene coated beads in a 4 × 5 well configuration. Beads can be washed with the aid of the Pentawash® II system (Abbott Laboratories) which is a hand-held aspirator/washer designed to wash 5 beads at a time. The enzyme/substrate used with the Abbott reagents is horseradish peroxidase/ϕ-phenylene-diamine (OPD) which results in a yellow complex in positive tests. Of the tests currently offered by EIA the anti HB_sAg test deserves special mention. This test utilizes the same polystyrene coated beads as other tests. However, antibody present in a patient's serum is detected using HB_sAg tagged with biotin, and avidin conjugated to horseradish peroxidase in order to increase the sensitivity of the test.[32,33] In addition to the tests listed in Table 1, an HTLV-III (human T-cell lymphotrophic virus III) antibody detection ELISA system is being developed based on the finding that both HTLV-III antigen and antibody are present in patients with acquired immune deficiency syndrome (AIDS).[34]

2. ELISA Processor M

The Behring ELISA Processor M (BEP) (Behring Diagnostics, LaJolla, Calif.) is an instrument designed for microtiter plate handling that incorporates an automated washer/dispenser, photometer, and printer into one unit. The washer/dispenser unit can be used in a manual or automatic mode and can be programmed to dispense variable volumes (25 to 275 *ℓ*) of reagents, to perform a variable number of washes (0 to 6), to give a variable soak time between washes (0 to 120 sec), or to aspirate only. Thus, the BEP can be used not only with the ENZYGNOST (Behring Diagnostics) system but also with any ELISA kit that uses a standard-sized microtiter plate with the wells as the solid phase. The photometer is an eight-channel fiber optic bichromatic system with the capability to read multiple wavelengths (405, 450, 492, 570, 650 mm). The microprocessor has 30 available programs, 10 of which are preprogrammed with the other 20 being user programmed. The instrument can process partial plates automatically. The comprehensive print-out includes flagging of abnormal results. The BEP system costs $21,500. The ENZGNOST tests use alkaline phosphatase and *p*-nitrophenylphosphate as the indicator system.

3. Photometer III

The Cordis Photometer III (Cordis Laboratories, Inc., Miami, Fla.) is a microprocessor-based reader and printer designed for use with Cordis ELISA kits. The photometer determines absorbance at a set wavelength of 405 nm and uses a flow-through cuvette for sample volumes of 0.7 m*ℓ* or less. There are software modules available for all Cordis assays. The list price of this system is $6000 with an additional $175 charge for each program module. Cordis also has available single or multiple disc transfer devices, and a single or multiple disc washing device that uses a foot-operated wash dispenser. A digital 405 Photometer II instrument is available for $1150 for manual use and data interpretation. The Cordis kits use antigen or antibody impregnated discs as the solid-phase and an alkaline phosphatase/*p*-nitrophenyl-phosphate enzyme/substrate system for antibody detection. A yellow end-product *p*-nitrophenol is formed in positive tests.

4. Whittaker M.A. Bioproducts

In examining Table 1, it should be noted that Whittaker M. A. Bioproducts (Walkersville, Md.) has not developed instrumentation for use with their ELISA kits but instead has taken the approach of developing tests which are compatible with many autoreaders, pipettors, and programmable readers and printers developed by other manufacturers. The ELISA kits utilize microtiter wells as the solid phase. The system is flexible in that Removawell® strips can be used to economically perform the exact number of tests required instead of using an entire microtiter tray. Most of the ELISA tests marketed by M.A. Bioproducts use the alkaline phosphatase/*p*-nitrophenylphosphate system for detection of antigen or antibody. The M.A. Bioproducts package inserts state that a Dynatech MR590 EIA plate reader (Dynatech Laboratories, Inc. Alexandria, Va.), Gilford Stasar® II (Gilford Co. Oberlin, Ohio), or equivalent should be used for reading the color of the end-products. An alternate choice would be the EL308 manual reader or EL310 automated reader and printer produced by Bio-Tek Instruments. Inc., Burlington, Vt.[35] The list price of the EL310 is $10,900. It uses an adjustable wavelength range of 380 to 690 nm.

Sever has recently reviewed the wide variety of Dynatech equipment available which includes hand-operated and automated equipment for adding reagents, diluting specimens, transferring and washing devices.[31] He also discussed the MR590 minireader (Dynatech Laboratories) and the automated MR600 microplate reader which can be used alone or in conjunction with an Apple® computer for data analysis. The MR600 can be programmed to read a 96-well microtiter plate in any format and can bused for hemagglutination inhibition and complement fixation tests as well as ELISA. It has an adjustable wavelength range of 410 to 630 nm. The MR600 with Apple® computer costs $13,500. Dynatech has programs available for use with M.A. Bioproducts' ELISA test kits.

Another system which is compatible with the M.A. ELISA tests is the Titertek Multiscan system (Flow Laboratories, McLean, Va.) which has the capability for full automation using a variety of diluters, washers, etc. developed by Titertek.

5. FP-900 EIA System

The FP-900 EIA system (Labsystems, Inc. Chicago, Ill.) is a modular system which can be used either as a manual system or can be fully automated. The fully automated system consists of an automated sampler, incubator, rinser, and analyzer. Each component can be purchased separately. While the total price for the entire system is approximately $24,000, the purchase price of the Analyzer (photometer) is only $8000. The FP-900 system is unique in that it is one of the few systems developed for both clinical chemistry and immunology. It utilizes a 9-well plastic cuvette block which the manufacturer states has a binding capacity 2½ times greater than conventional microtiter plate wells, enhancing control of assay sensitivity. The analyzer uses a vertical measurement principle, has a wavelength range from 340 to 850 nm, and can measure 9 samples simultaneously. The FP901 analyzer can be interfaced with an external computer. A nine-channel pipette can be used as a rapid manual washing device. Labsystems also markets the Uniskan of which there are three types; a small battery-operated portable photometer printer, a manually operated system and an automated reader-printer. The EIA procedures (Table 1) for serodiagnosis of infectious diseases utilize alkaline phosphatase/*p*-nitrophenylphosphate for detection of antibodies.

6. LB-321 Automated ELISA System

The LB-321 ELISA system (Litton Bionetics, Charleston, S.C.) includes a reader, computer, and printer in one unit as well as a benchtop 37°C incubator-shaker and a Magnetic Transfer Device for transfer of the solid-phase (metal beads). The LB-300 system consists of the first unit described and uses 8 light-emitting diodes to give a constant monochromatic light source of 692 nm. The list price of the LB-300 system is $14,300 while the cost of the LB-321 is $15,945. An automated dilutor is also available, costing approximately $2400. The system is designed for use with the Bio-enzabead test kits produced by Litton (Table 1). The same principle of heterogeneous EIA is used by Litton with an unusual feature, that of using polycarbonate coated metal beads as the solid phase. The use of the Magnetic Transfer Device (Litton Bionetics) enables one to transfer up to 96 metal beads simultaneously from one reaction mixture to another. The Bio-enzabead kits use horseradish peroxidase/2,2′ azino-di[3-ethylbenzthiazoline sulfonate (6)] (ABTS) as the detector system. A positive test is evidenced by the development of an emerald green color due to oxidation of the ABTS substrate. In addition to the tests listed in Table 1, Litton is also developing an HTLV-III antibody test.

7. Ortho® Microwell Test System

The Ortho® Microwell Test System (Ortho Diagnostic Systems, Raritan, N.J.) includes a number of ELISA test kits (Table 1) as well as the Ortho® Microwell reader and printer which is fully computerized and uses data reduction cartridges. The photometer is set to read at a wavelength of 490 nm using a tungsten halogen light source. An Ortho® Microwell washer capable of washing 12 wells at a time is also available. The reader has a list price of $4800, but can be obtained free based on test volume.

The ELISA assay kits contain Removawell® strips (Ortho) with the individual wells coated with an appropriate antigen or antibody. The enzyme/substrate system used in horseradish peroxidase/ϕ-phenylenediamine. In addition to the tests listed in Table 1, Ortho is also working on a microwell ELISA system for the serologic detection of HTLV-III antibodies. Their tests do not have to be read with the Microwell reader (Ortho) but can be analyzed on any spectrophotometer capable of reading at 490 nm.

III. SUMMARY

Systems which are manually operated except for photometric or fluorometric readers, semiautomated, and entirely automated systems have been developed for use in the clinical serology laboratory. These systems are adaptable to, or were developed for, the serodiagnosis of infectious diseases. Some instruments may be used only with their respective manufacturer's test kits while others are usable with several reagent systems. In several instances, automation has resulted in simple, rapid, cost-effective testing for microbial antigens or antibodies produced against them.

EIA procedures and semiautomated FIA procedures will no doubt continue to replace RIA techniques due to the advantages enumerated above. In addition, sensitivity of EIA and FIA procedures should increase. The use of higher energy substrates, such as fluorescent and luminescent substrates and thermometric analytes,[25] as well as the increased use of biotin-avidin or plant lectins to increase the binding of enzyme conjugate-substrates will aid in development of test systems capable of identifying a wider array of microbial antigens and antibodies. More fluorometric instruments for serology will be developed as fluorophores yielding greater sensitivity and stability are discovered and developed. These instruments may use standard fluorometers or may be developed to include nitrogen laser methodology in which a high intensity, monochromatic light beam excites the fluorophore.[17] Further development of fluorometric techniques such as fluorescence polarization and time-resolved fluoroimmunoassay will also increase the desirability of these methods.

Finally, there will continue to be more and more emphasis placed on antigen detection as a means of rapid diagnosis. However, this will not entirely replace standard serologic testing. Development of highly specific and sensitive monoclonal antibodies will increase detection of both antigens and specific immunoglobulins. New tests, such as anti-HTLV-III, will assume increasing importance. The number of tests able to detect IgM antibodies for early diagnosis of primary or reactivation disease will increase. In addition, automation will make possible the use of the "continuous titer" concept,[36] in which multiple specimens may be collected and tested over a short period of time, potentially decreasing the time necessary to demonstrate a significant change in titer to as short as 2 to 3 days. Thus, antibody measurement will continue to play a significant role in the serodiagnosis of many infectious diseases.

REFERENCES

1. Walsh, J. H., Yalow, R. S., and Berson, S. A., Detection of Australia antigen and antibody by means of radioimmunoassay techniques, *J. Infect. Dis.*, 121, 550, 1970.
2. Catt, K. and Tregear, G. W., Solid phase radioimmunoassay in antibody-coated tubes, *Science*, 158, 1570, 1967.
3. Overby, L. R. and Mushahwar, I. K., Radioimmunoassays, in *Rapid Diagnosis in Infectious Disease*, Rytel, M. W., Ed., CRC Press, Boca Raton, Fla., 1979, 39.
4. Daugharty, H. and Ziegler, D. W., Radioimmunoassay in viral diagnosis, in *Comparative Diagnosis of Viral Disease*, Vol. 2, Kurstak, B. E. and Kurstak, C., Eds., Academic Press, New York, 1977, 459.
5. Schinski, V. D., Clutter, W. C., and Murrell, K. D., Enzyme and ^{125}I-labeled anti-immunoglobulin assays in the immunodiagnosis of schistosomiasis, *Am. J. Trop. Med. Hyg.*, 25, 824, 1976.
6. Straus, E., Wu, N., Quraishi, M. A. H., and Levine, S., Clinical applications of the radioimmunoassay of secretory tuberculin, *Proc. Natl. Acad. Sci. U.S.A.*, 78, 3214, 1981.
7. Yalow, R. S., Radioimmunoassay, *Clin. Immunol. Newsl.*, 5, 23, 1984.
8. Frisch, C. and Whiteside, T. L., A comparison of the FIAX and hemagglutination inhibition tests for IgG antibodies to rubella virus, *J. Clin. Lab. Automat.*, 1, 33, 1981.
9. Segar, J. E., Smith, T. F., and Ilstrup, D. M., Evaluation of the FIAX test for the detection of antibodies to herpes simplex virus and cytomegalovirus, *Am. J. Clin. Pathol.*, 75, 387, 1981.
10. Friedman, H. M., Tustin, N. B., Hitchings, M. M., and Plotkin, S. A., Comparison of complement fixation and fluorescent immunoassay (FIAX) for measuring antibodies to cytomegalovirus and herpes simplex virus, *Am. J. Clin. Pathol.*, 76, 305, 1981.
11. Hyde, B., Burgett, M. W., and Maggio, E. T., A solid-phase fluoroimmunoassay for the quantitation of antibodies to *Toxoplasma gondii*, *Clin. Chim. Acta*, 103, 393, 1980.
12. Walls, K. W. and Barnhart, E. R., Titration of human serum antibodies to *Toxoplasma gondii* with a simple fluorometric assay, *J. Clin. Microbiol.*, 7, 234, 1978.
13. Walls, K. W., Immunofluorescence and immunoenzymatic assays for diagnosing parasitic infections, in *Rapid Methods and Automation in Microbiology*, Tilton, R. C., Ed., American Society for Microbiology, Washington, D.C., 1981, 134.
14. Taylor, R. G. and Perez, T. R., Serology of amoebiasis using the FIAX system, *Arch. Invest. Med. (Mexico)*, 9, 363, 1978.
15. Walls, K. W. and Wilson, M., The use of the solid phase indirect immunofluorescent assay (FIAX) in the serodiagnosis of amoebiasis, *Ann. N.Y. Acad. Sci.*, 420, 422, 1983.
16. Stevens, R. W., Recent advances in agglutination and fluoroimmunoassays for the diagnosis of bacterial infections, in *Rapid Methods and Automation in Microbiology*, Tilton, R. C., Ed., American Society for Microbiology, Washington, D.C., 1981, 121.
17. Soini, E. and Hemmila, I., Fluoroimmunoassay: present status and key problems, *Clin. Chem.*, 25, 353, 1979.
18. Dandliker, W. B., Shapiro, J. H., Meduski, J. W., Alonso, R., Feigen, G. A., and Hamrick, J. R., Application of fluorescence polarization to the antigen antibody reaction, *Immunochemistry*, 1, 165, 1964.

19. Dandliker, W. B., Kelly, R. J., Dandliker, J., Farquhar, J., and Levin, J., Fluorescence polarization immunoassay, theory and experimental method, *Immunochemistry*, 10, 319, 1973.
20. Halonen, P., Meurman, O., Lovgren, T., Hemmila, I., and Soini, E., Detection of viral antigens by time-resolved fluoroimmunoassay, *Curr. Topics Microbiol.*, 104, 133, 1983.
21. Muerman, O. H., Hemmila, I. A., Lovgren, T. N. E., and Halonen, P. E., Time-resolved fluoroimmunoassay: a new test for rubella antibodies, *J. Clin. Microbiol.*, 16, 920, 1982.
22. Gross, M., personal communication, 1984.
23. Burd, J. F., Wong, R. C., Feeney, J. E., Carrico, R. J., and Boguslaski, R. C., Homogeneous reactant-labeled fluorescent immunoassay for therapeutic drugs exemplified by gentamicin determination in human serum, *Clin. Chem.*, 23, 1402, 1977.
24. Avraneas, S., Enzyme immunoassays and related techniques: development and limitations, in *Current Topics in Microbiology: New Developments in Diagnostic Virology*, Vol. 104, Bachmann, P. A., Ed., Springer-Verlag, New York, 1983, 93.
25. Oellerich, M., Enzyme immunoassays in clinical chemistry: current status and trends, *J. Clin. Chem. Clin. Biochem.*, 18, 197, 1980.
26. Schall, R. F., Jr. and Fraser, A. S., Enzyme immunoassays: a technical review, *Med. Device and Diagn. Ind.*, 5, 66, 1983.
27. Yolken, R. H., Enzyme immunoassays for the detection of microbial antigens in body fluids, *Clin. Immunol. News.*, 3, 25, 1982.
28. Yolken, R. H., Leister, F., Whitcomb, L., Davis, D., and Mears, M. J., Enzyme immunoassays for the diagnosis of viral infections, *Ann. N.Y. Acad. Sci.*, 420, 381, 1983.
29. Yolken, R. H., Enzyme immunoassays for the detection of infectious antigens in body fluids: current limitations and future prospects, *Rev. Infect. Dis.*, 4, 35, 1982.
30. Parksinson, A. J., Scott, E. N., and Muchmore, H. G., Comparison of a solid-phase enzyme immunoassay and a radioimmunoassay using a double-labeled (enzyme-^{125}I) antibody, *J. Clin. Lab. Automat.*, 2, 177, 1982.
31. Sever, J. L., Automated systems in viral diagnosis, in *Current Topics in Microbiology: New Developments in Diagnostic Virology*, Vol. 104, Bachmann, P. A., Ed., Springer-Verlag, New York, 1983, 57.
32. Bager, E. A. and Wilchek, M., The use of the avidin-biotin complex as a tool in molecular biology, *Methods Biochem. Anal.*, 26, 1, 1979.
33. Kendall, C., Ionescu-Matiu, I., and Dreesman, G. R., Utilization of the biotin/avidin system to amplify the sensitivity of the enzyme-linked immunosorbent assay (ELISA), *J. Immunol. Methods*, 56, 329, 1983.
34. Gallo, R. C., Salahuddin, S. Z., Popovic, M., Shearer, G. M., Kaplan, M., Haynes, B. F., Palker, T. J., Redfield, R., Oleske, J., Safai, B., White, G., Foster, P., and Markham, P. D., Frequent detection and isolation of cytopathic retroviruses (HTLV-III) from patients with AIDS and at risk for AIDS, *Science*, 224, 500, 1984.
35. Gibson, S., personal communication, 1984.
36. Webert, D. W. and Bech-Nielsen, B. V., Detection of early "significant" antibody responses in paired sera by the use of an automated bovine parainfluenza-3 hemagglutination-inhibition system, *Am. J. Clin. Pathol.*, 70, 679, 1978.

Chromatographic Methods

Chapter 12

GAS CHROMATOGRAPHY AND MASS SPECTROMETRY

Lennart Larsson

TABLE OF CONTENTS

I. INTRODUCTION

Gas chromatography (GC) is used in clinical microbiological laboratories mainly in the identification of strict anaerobic bacteria by analyzing short-chain fatty acids excreted by the anaerobes into a liquid culture medium. GC studies on constituents of microorganisms can also be used in differential diagnosis; for instance, in laboratory identification of nonfermentative and slow-growing bacteria by means of cellular fatty acid analysis. Promising results have also been reported on the use of GC in the rapid diagnosis of infectious diseases by direct analysis of metabolites and specific microbial constituents in clinical specimens.

When combined with mass spectrometry (MS), the analytical power of GC is considerably enhanced. In contrast to GC used on its own, GC/MS enables the separated compounds to be unequivocally identified.[1] In addition, when adapted to selected ion monitoring (SIM), MS can be used as a detector of high selectivity and sensitivity. Both of these features make GC/MS superior to merely GC in analyses of microbial metabolites and constituents.

This chapter is intended as a brief review of the GC/MS technique and its application in diagnostic microbiology. Recent technological advances are discussed, including computerization, automation, and the development of cheaper and more compact GC/MS instruments.

II. BASIC PRINCIPLES OF GC/MS

A. Gas Chromatography

GC is an excellent method for separating compounds of moderate molecular weight. A major breakthrough in GC technology was achieved with the introduction of open tubular (capillary) columns practical for use in routine laboratory work and capillary injectors for reproducible sample introduction. Compared with packed columns (typically 2 m in length and 2 to 4 mm inner diameter), capillary columns (typically 25 or 50 m in length and with inner diameters of 0.2 or 0.35 mm) possess superior separation qualities. Fused silica open tubular (FSOT) columns can be coated with a stationary phase of either low or high polarity[2] and, unlike glass capillaries, are both flexible and durable.

Owing to the low sample capacity of narrow-bore capillaries, however, special techniques are necessary for sample introduction. In the *split injection* technique, the sample vapors are divided in two unequal parts immediately after injection, only the smaller part being allowed to enter the column for separation and identification. *Splitless injection*[3] implies purging off part of the solvent shortly after injection and is preferred when limited amounts of material are available for analysis. A third technique, *on-column injection,* usually gives good results (high sensitivity and equally effective with compounds of high or low boiling point) and has recently been developed to be connectable to an automatic injection unit.

GC detectors exist with a variety of performance characteristics. The *thermal conductivity* detector is widely applicable but has a comparatively low sensitivity. The *flame ionization* detector responds to all organic molecules with a typical lowest detectable amount of around 1 ng, and is frequently used in the analysis of microorganisms. The *electron capture* detector responds with extremely high sensitivity to electrophilic compounds such as molecules possessing one or more halogen atoms, but is insensitive to nonelectrophilic compounds. Finally MS adapted to *SIM* constitutes a GC detector which is universally applicable, yet very selective and sensitive; this detector will be described in some detail in the following.

B. Gas Chromatography/Mass Spectrometry

1. The Interface

The GC/MS interface provides a link between the GC column which operates at atmospheric pressure, and the MS which frequently operates under high vacuum. It concentrates the sample molecules prior to their entering the MS, by selectively removing carrier gas molecules. An all-glass jet separator is normally used for packed GC columns, whereas for narrow-bore capillary columns, operating at much lower carrier gas flow rates, the total effluent from the column can be transported by a transfer line to the ion source. This technique ensures maximum sample transfer and low dead volume, and is particularly well-suited for use with the flexible FSOT columns which can be directly interfaced to the ion source. For wide-bore capillary columns, a so-called open-split interface is often recommended.

2. The Ion Source

The sample molecules are transferred from the interface to the ion source of the MS and ionized, either by electron bombardment (electron impact, EI), or by ion/molecule collisions (chemical ionization, CI).[1] In EI, typically using an electron energy of 70 eV and a pressure of 10^{-7} torr, the parent ions initially formed largely decompose to a variety of radicals and ions of which the latter are separated and subsequently analyzed (see below). Vacuum is achieved by high-efficiency turbomolecular or diffusion pumping systems. When using CI, the pressure is increased to approx. 1 torr and a reactant gas in introduced in the ion source. This reactant gas is ionized by electrons, the resulting ions being used to ionize further the sample molecules. In CI the fragmentation of the parent ions is greatly reduced compared with EI, and CI mass spectra frequently show abundant ions at $(M + 1)^+$ or $(M - 1)^+$. Switching between EI and CI can be done almost instantaneously with modern instruments.

Both in EI and CI, positive ions are predominantly used for study, though a small portion of the parent ions will be negatively charged. During recent years, negative ion (NI) MS technology, utilizing either EI or CI,[4] has developed considerably, and for electrophilic compounds a very high sensitivity can be obtained. Thus, NI MS is suitable for trace analysis of the same type of molecules as used in the electron capture detector.

3. Separation of Ions

Two major methods for ion separation are in current use.[1] In *magnetic sector* instruments, the ions are accelerated by an electrostatic voltage to a magnetic field and there deflected along a curvature. By continuously scanning the magnetic field strength, or the accelerating voltage, ions of different mass-to-charge ratios are consecutively allowed to pass through an exit slit for measurement. In *quadrupole* instruments, four electrodes placed at the vertices of a square are used. A pair of electrodes, at the opposite angles, are connected and supplied with a DC voltage and a high-frequency AC voltage. The masses are scanned by sweeping the voltages keeping the relationship between the DC and AC values constant. Modern magnetic sector and quadrupole MS instruments are capable of scanning as fast as a few tenths of a second per decade, a feature which is particularly important when using capillary GC columns from which compounds may elute very rapidly. The results of the ion measurements can be represented as a graph, a mass spectrum, which may be regarded as a characteristic "fingerprint" of the specific compound studied. With magnetic sector instruments, it is possible to achieve higher resolution (i.e., superior ability to separate ions of similar mass-to-charge ratios) than with quadrupole instruments.

When adapted for SIM, the MS is focused at only one ion, or at a limited number of ions, rather than being allowed to scan continuously over a mass range.[5] Thus, since much more time is spent studying each ion in SIM, considerably higher sensitivity is possible. Used in this way, the MS may be regarded as a GC detector. If the ions focused at are specific for the compounds under study, this detector will possess a very high degree of selectivity. Its sensitivity depends upon the abundance of ions at which it is focused, but often exceeds that of a flame ionization detector by a factor ranging from 100 to 1000. Most modern GC/MS instruments are capable of monitoring several ions "simultaneously" which means that more data can be obtained, though this also results in a decrease of sensitivity compared with monitoring a single ion. When using SIM, CI MS, resulting in the formation mainly of ions of masses (M + 1) or (M − 1), is often preferred to EI for reasons of sensitivity. For ultrasensitive analysis of microbial constituents or metabolites using NI MS, derivatization to halogen-containing molecules is required.

4. Detection of Ions

In almost all commercially available mass spectrometers, electron multipliers are used for ion detection and amplification of the resulting signals.

5. Computerization and Automation

Magnetic sector and quadrupole GC/MS instruments are available with facilities for automatic sample injection, and adaptable to different, preprogrammed GC conditions as well as to conditions for ionization and ion scanning, computerized evaluation of results, etc. These features enhance the capacity of GC/MS as an instrument that can operate even beyond laboratory working hours. In computerized GC/MS, ions are typically scanned repetitively and the data stored in the computer memory. After each chromatographic run, mass spectra from each of the scannings can be displayed on a screen or printed out. From repetitive scanning data it is possible to display "extracted ion current profiles", i.e., chromatograms in which only ions of selected mass-to-charge ratios are included. Further, in a "library search" function, mass spectra of the unknown compounds are compared with reference spectra, stored in a data bank. Suggestions for molecular structures of the unknown components, with probabilities indicated, are automatically listed. Such data banks, containing several thousands of reference spectra, are available as computer software and greatly simplify the evaluation of mass spectra.

6. Mass Selective Detectors

Recently, compact MS instruments, designed to allow their connection to almost any capillary GC, have become commercially available. These so-called mass selective detectors (MSD) are largely computerized and considerably less expensive than ordinary GC/MS instruments. They can be used both for identifying GC eluates by uptake of mass spectra, and for selective detection in combination with SIM. Only ionization with EI is possible with the instruments developed so far.

The Hewlett-Packard (HP 5970) MSD instrument provides a choice of an open-split or direct interface between the capillary column and the ion source. The mass filter consists of a hyperbolic quadrupole, and ions within a mass range of 10 to 800 amu (atomic mass units) can be analyzed. The instrument can be focused at up to 20 ions "simultaneously" using SIM. The Finnegan "ion trap detector" (ITD) model also utilizes a quadrupole and is very compact owing to the ion source and mass filter having been combined in a single unit. As with the HP 5970 instrument, only EI ionization is possible, and the spectrometer has a mass range of 10 to 650 amu. Up to sixteen ions can be monitored in the SIM mode.

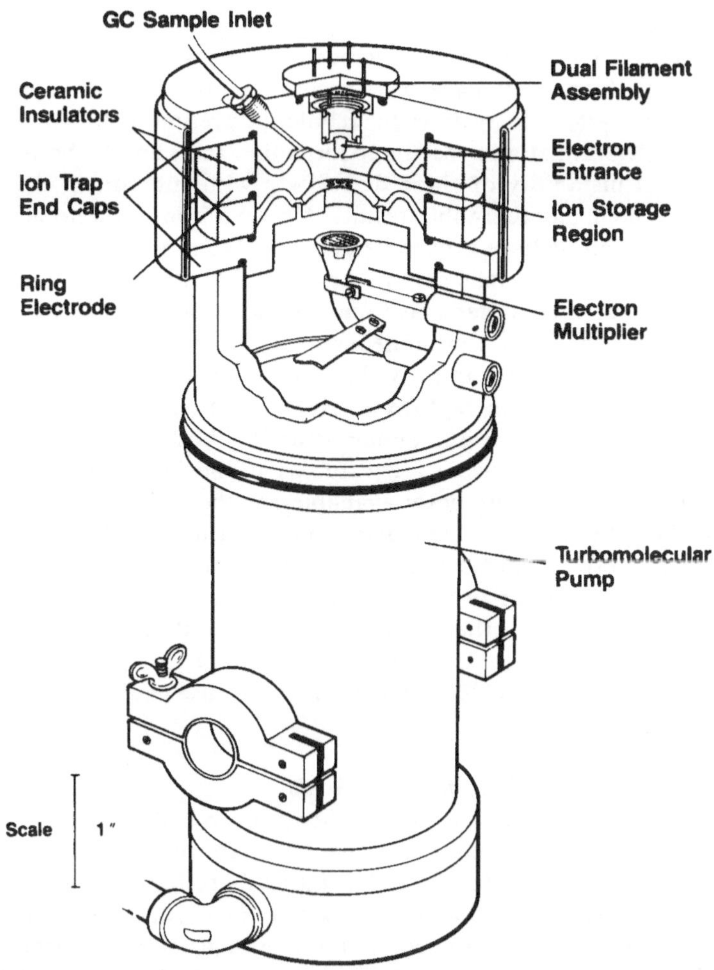

FIGURE 1. Schematic diagram of the "ion trap" detector. (By kind permission of Finnegan MAT Ltd., England.)

Although MSD instruments seem very attractive for use in routine diagnostic work, it should be stressed that experience with these instruments in clinical laboratories is still very limited. A major disadvantage of the MSD instruments currently available is their inability to utilize ionization with CI and NI, which restricts their sensitivity and selectivity.

III. APPLICATIONS IN CLINICAL MICROBIOLOGY

When employing GC and GC/MS analysis for obtaining taxonomic and diagnostic information in microbiology, one or more of the following types of samples are used for study:

1. Pure microbial isolates for analysis of cellular constituents
2. Spent culture medium for analysis of metabolic end-products
3. Clinical samples for analysis of metabolites or specific cellular constituents

Each of these areas of application will now be discussed.

A. Cellular Constituents

The major cellular constituents studied include fatty acids, amino acids and carbohydrates, of which the fatty acids have been predominantly used for diagnostic purposes. Reasons for this fatty acid preference include the present methods' complicated and time-consuming derivatization procedures for amino acids and carbohydrates, and sometimes a chemical instability of these derivatives with subsequent difficulties of achieving reproducible chromatographic results. However, both amino acid and carbohydrate analysis of bacteria have proved valuable in taxonomic studies. For example the demonstration of diaminopimelic acid in *Legionella* spp. was helpful in clarifying the taxonomic position of these organisms,[6] and characterization of bacteria[7-9] and of bacterial lipopolysaccharides[10] by simultaneous analysis of fatty acids and carbohydrate monomers has been reported. Improved analytical techniques involving capillary column GC and sensitive (electron capture or SIM) detection, have recently been developed for use with carbohydrates and amino acids,[11-13] both of which are classes of compounds that are likely to be increasingly studied in the future. Reproducible and generally accepted methods are presently available for the hydrolysis, derivatization and analysis of cellular fatty acids. These acids, both free and lipid bound, can easily be converted to methyl esters which are so chemically stable that preparations can be stored for several months without any detectable changes occurring in the corresponding chromatographic profiles. Further, results from bacterial fatty acid analyses usually tally with the present taxonomic designations. Methods for preparation and GC analysis of methyl esters of such acids, suitable for routine use in differential diagnosis, will be discussed in the following.

1. Preparation and Analysis of Cellular Fatty Acids

Typically, lipids are not extracted from microbes prior to hydrolysis. Two principles for conversion of cellular fatty acids to methyl esters are predominantly used, each requiring a few mg of cells when using flame ionization detection. One method involves heating of the microbes in acid methanol (typically at 80°C for 1 to 18 hr), by which depolymerization and methylation are achieved in one single step (i.e., transesterification), after which the methyl esters are extracted from the hydrolysates. This method releases the amide-bound hydroxy acids quantitatively, and converts carbohydrates to methyl glycosides, but has the disadvantage that it degrades cyclopropane-substituted acids. In an alternative method the microbes are heated in methanol containing sodium hydroxide solution (typically at 80 to 100°C for 30 min) for saponification of the lipids. After acidification of the samples, the liberated fatty acids are methylated by adding methanol and heating for a few minutes, after which the esters are extracted and analyzed. The alkaline saponification method does not degrade cyclopropane-substituted fatty acids, but leads to an incomplete release of amide-bound hydroxy acids, and degrades carbohydrates. Use of 15% NaOH in methanol-water (1/1, v/v) for saponification has been recommended, resulting in a release of hydroxy acids of at least 50%.[14]

For simultaneous analysis of fatty acids and carbohydrates, acid methanolysis followed by derivatization of the resulting methyl glycosides using trifluoroacetic anhydride is a useful technique.[10,15] This method provides more informative GC results, and is particularly valuable in cases when chromatograms with an insufficient degree of specificity are obtained from the analysis of fatty acids alone.

As not all of the fatty acid, amino acid and carbohydrate derivatives which may occur simultaneously in a bacterial isolate can be separated on a packed GC column, FSOT columns are increasingly used. Such columns, coated with a nonpolar stationary phase, are capable, for example, of separating most bacterial fatty acids esters.[16] In

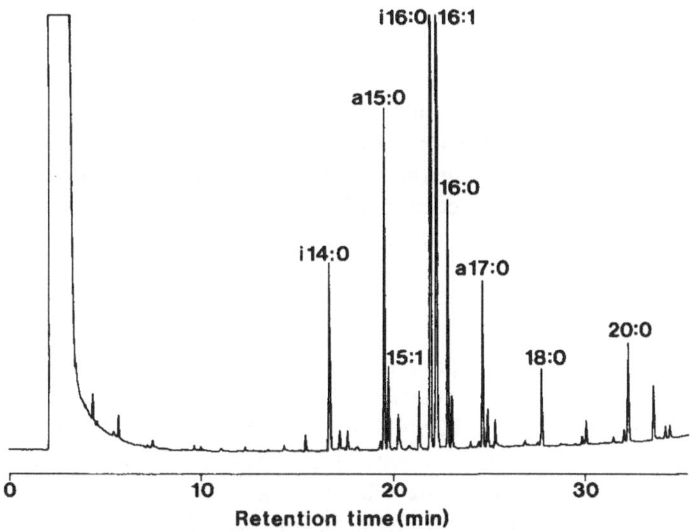

FIGURE 2. Cellular fatty acid chromatogram of *Legionella pneumophila* serogroup 1. The designations of the peaks refer to number of carbon atoms and double bonds; *a* and *i* indicate anteiso- and iso-branched acids, respectively.

addition, when using unsplit (splitless or on-column) injection[17] onto FSOT columns, an improved sensitivity compared with that of packed columns is achieved, because with capillary columns the electronic attenuation of the detector signal used is lower and the peaks narrower and higher. Prior to GC analysis, particularly when using the alkaline saponification method, care must be taken to ensure the removal of traces of free fatty acids possibly present in the final extract due to incomplete methylation of the liberated acids. Unless this is done (e.g., by extraction with an alkaline buffer solution), hydroxy acid esters or free fatty alcohols may be adsorbed in the column.[18,19]

2. Diagnostic Implications

The cellular fatty acid distributions have been evaluated in numerous microbial species. Diagnostic use of GC analysis would be attractive in the case of microbes where alternative identification procedures are uncertain, laborious or time-consuming. Glucose nonfermentors, such as *Achromobacter, Alcaligenes* and *Pseudomonas* spp., constitute one group of bacteria which has been relatively extensively studied by GC.[20] Their fatty acid profiles are often either species-specific or characteristic for a small group of species belonging to these genera. Further, all species of *Legionella* can be distinguished from other Gram-negative bacteria by characteristically high proportions of branched-chain fatty acids and the presence of only minor amounts of hydroxy acids. In addition, many *Legionella* species can be differentiated by consistent, quantitative differences in their fatty acid profiles.[21] Members of the family *Neisseriaceae*, for which definite identification is often difficult using conventional techniques, constitutes another group of organisms where this GC approach may be helpful.[22] As several species of the genera mentioned contain cyclopropane-substituted acids, it is important to avoid acid hydrolysis and consequent degradation of these acids. Excellent results have been reported for *Mycobacterium*. By conventional methods, species identification of mycobacteria may take several weeks, whereas by GC, this can often be accomplished within a few hours. Methods employed in this connection mainly include fatty acid analysis,[19,23-25] although also use of combined analysis of fatty acids and carbohydrates (the latter analyzed as trifluoroacetylated methyl glycosides) on a capillary column has been reported.[15] The interlaboratory reproducibility of fatty acid profiles for mycobacteria has been partly evaluated.[24,25] and the encouraging results obtained indicate the possibility of constructing a "library" of reference chromatograms. In order to ascertain separation of secondary alcohols (2-octadecanol and 2-eicosanol), it is important to use a properly deactivated GC system to prevent these alcohols from being absorbed in the column. The alcohols are released from dicarboxyl mycolic acids upon hydrolysis and constitute important chemical markers in species differentiation.[19]

B. Metabolic End-Products

The most widely employed GC approach in clinical microbiological laboratories is in the identification of anaerobic bacteria. Many anaerobes produce relatively large amounts of short-chain fatty acids. These acids are easily extracted from broth cultures after acidification using an organic solvent. Nonvolatile acids, such as lactic and succinic acids, are converted to methyl esters by adding methanol and a catalyst such as BF_3 or BCl_3 to the cultures, followed by heating for 5 to 10 min, after which the esters are extracted and analyzed.[26] Alternatively, broth cultures may be directly injected into the GC column. Relatively recently, the use of head-space GC (HSGC) was described.[27-29] In automatic HSGC analysis, spent culture medium (typically 1 m*l*) is transferred to glass ampoules designed to fit in the heated turntable connected to the gas chromatograph. For analysis of volatile acids it is essential to first acidify the specimens, e.g., by adding sulfuric acid[27,28] whereas for volatile amines, increasing the pH by adding, i.e., solid sodium hydroxide is useful.[29] HSGC has the advantage that it makes extraction procedures unnecessary which means saving of time and labor, and that no solvent peak is present in the chromatograms (when a flame ionization detector is used). This latter feature makes rapidly eluting metabolites, such as short-chain alcohols, readily detectable. Use of capillary columns both in HSGC and in analysis of the short-chain fatty acids as butyl esters after extraction has been described.[27,28,30]

Using GC alone, strictly anaerobic microbes can often be identified as to genus, but less often as to species. The *Anaerobic Laboratory Manual* [26] (Virginia Polytechnic Institute, Blacksburg, Va.) contains chromatograms, obtained by packed column GC, of virtually all clinically relevant anaerobic bacteria. No comprehensive data are available from the use of improved GC techniques such as those mentioned above, although such data would most probably provide a higher degree of specificity. With few exceptions, facultative anaerobes and aerobes produce only small amounts of short-chain fatty acids which are of limited diagnostic value. Examples of metabolic products other than fatty acids which have been studied (and which are perhaps chiefly of interest as curiosities) are the identification of 2-aminoacetophenone as being responsible for the typical odor of *Pseudomonas aeruginosa*,[31] and *p*-cresol giving the characteristic odor of *Clostridium difficile*.[32] However, successful results have been reported from analyses of both acid, neutral and alkaline metabolic end-products of a variety of both aerobic and anaerobic species, after derivatization of these metabolites to halogen-containing compounds and analyses using an electron capture detector.[33] Automated HSGC analysis can be used for rapid differentiation between several Gram-negative aerobic bacteria by analysis of volatile products from their degradation of glucose,[34] and has also proved valuable in the diagnosis of bacteremia by analysis of blood cultures.[35]

C. Clinical Samples

Both relatively simple and highly sophisticated laboratory procedures are being used in direct GC studies on clinical samples for the rapid diagnosis of infectious diseases. This application of GC analysis involves demonstration, in clinical specimens, of specific chemical compounds which can be used as marker molecules of the infections. These chemical markers may either be constituents or products of the causative microbes, or compounds originating from a specific body response to a particular disease. In the following a selection of such markers will be discussed, including methods for their detection, and their diagnostic implications.

1. Short-Chain Fatty Acids

The presence of appreciable amounts of short-chain fatty acids in, for example, pus, sputum, and pleural and joint aspirates, strongly suggests that a given infection is caused by anaerobic bacteria. Although both false-positive and false-negative results may occur, and the resulting chromatograms fail to identify either the species or genus of the causative bacteria, nonetheless this diagnostic approach can be very useful for rapid demonstration of anaerobes in clinical samples such as those mentioned above.[36,37] The fatty acids are generally extracted from the specimens after acidification and are analyzed by GC using flame ionization detection. Alternatively, clinical samples can be directly injected onto the column, or analyzed using HSGC.

Since specimens from patients with such diseases as bacterial meningitis, peritonitis and septic arthritis often contain lactic and succinic acids in concentrations above normal values, these compounds may serve as markers for differentiating between bacterial and non-bacterial inflammation. False-negative results have been reported for gonococcal infections and for some cases of bacterial meningitis in early stages of infection, however.[38,39] Both of the acids mentioned are easily converted to methyl esters and analyzed using flame ionization detection.

Isocaproic acid is often found in stools in conjunction with toxin from *Clostridium difficile,* an organism which has been claimed as a causative agent of pseudomembranous colitis. The reliability of GC results seems to be related to the detector used: flame ionization detection of free isocaproic acid has been found to yield unsatisfactory results unless the extracts were concentrated prior to analysis;[40,41] whereas, when using electron capture detection of the same acid converted to its trichloroetyl ester, excellent correlations with *C. difficile* toxin findings were obtained.[42] False-positive results have been reported where isocaproic acid was detected in stools from which *C. bifermentans* and *C. sordellii,* but not *C. difficile,* were later isolated.

2. Arabinitol

D-Arabinitol is product of many fungi. It has been suggested that, after derivatization to trimethylsilyl ether, it may be used as a marker for disseminated *Candida* infection by GC analysis of serum. The methods hitherto employed include total analysis of the combined D- and L-arabinitol content, using either flame ionization[43] or SIM[44] detection. As appreciable amounts of the latter stereoisomer are present in normal serum, diagnosis has so far been based upon measurements of the increase in the combined concentration of D- and L-arabinitol.

Use of a chiral stationary phase in the GC column would enable separation of the two isomers and make improved analytical results possible to achieve. Access to a rapid diagnostic method for disseminated *Candida* infection is important, for example, in cancer hospitals where the patients may become infected with fungi. By GC analysis, such diagnosis can be made within a few hours although high arabinitol concentrations were recently reported also in sera from patients with sarcoidosis.[43]

3. Amino Acids and Carbohydrates

Sensitive GC and GC/MS assays have been developed for the analysis of specific bacterial amino acids and carbohydrates. The possibility exists that such components can be demonstrated directly in clinical samples. Using EI SIM detection, muramic acid — which is present in the peptidoglycan moiety of all bacteria — was demonstrated in tissues and joints of rats in which arthritis had been induced by the injection of streptococcal cell wall fragments, but could not be detected in samples from patients with rheumatoid arthritis.[45,46] Other potential chemical markers for the detection of bacteria and bacterial fragments in clinical material are D-amino acids[47] and 2-keto-deoxyoctonic and diaminopimelic acids, the latter two compounds being present mainly in Gram-negative bacteria. For utmost sensitivity, NI MS is the recommended detection mode.

4. Cellular Fatty Acids

β-Hydroxymyristic acid constitutes a part of the lipopolysaccharide fraction in Gram-negative bacteria and has been suggested as a marker for septicemia, though the EI SIM analyses were obstructed by interfering compounds in serum.[48] Again, alternative GC/MS conditions, such as NI MS analysis of halogenated derivatives, would provide superior sensitivity and selectivity.

Specific mycobacterial fatty acids have been demonstrated in sputum[49-51] and cerebrospinal fluid[52] specimens, as well as in their short-term incubated cultures. This approach implies the possibility of making positive diagnosis of tuberculosis and other mycobacterial infections within a few hours of collecting samples. Both tuberculostearic (10-methyloctadecanoic) and C-32 mycocerosic (2,4,6,8-tetramethyloctacosanoic) acids have been used as marker substances in capillary GC separation using both EI and CI SIM detection. The specimens are hydrolyzed and the liberated fatty acids methylated, sometimes purified using thin-layer chromatography, before being subjected to analysis.

Tuberculostearic acid can be detected in sputa which are negative for acid-fast rods in direct microscopy studies. It has not been found in samples from healthy patients, whether or not vaccinated against tuberculosis. The technique is highly promising as a tool for the rapid diagnosis of mycobacterial infections, offering several weeks of reduction in the time required for the diagnosis as compared with cultivation methods. Particularly in the case of tuberculous meningitis, access to the method may be of utmost importance for a successful treatment of the patient.

5. Compounds Resulting from Specific Body Response

The chemical composition (with regard to free fatty acids, amines, and alcohols) of certain body fluids such as serum, joint fluid, and cerebrospinal fluid, may sometimes be related to specific infectious agents. After selective extraction of these metabolites from clinical samples, derivatization is performed using halogenated reagents, and analysis using electron capture detection. This technique has been reported to yield useful differential diagnostic information for a number of infectious agents.[53] For example, chromatograms from analysis of short-chain acids in sera from patients with Rocky Mountain spotted fever, *Neisseria meningitidis* infections, and infections due to measles and rubella, were different.[54] Also, analysis of cerebrospinal fluid specimens from patients with meningitis due, e.g., to tuberculous and fungal infections produced different chromatographic results. The technique seems promising and may gain widespread acceptance in the future although this will require the development of simplified methodology regarding extraction, derivatization, etc.

IV. CONCLUSIONS

GC and GC/MS have several areas of application in clinical microbiology. GC is an

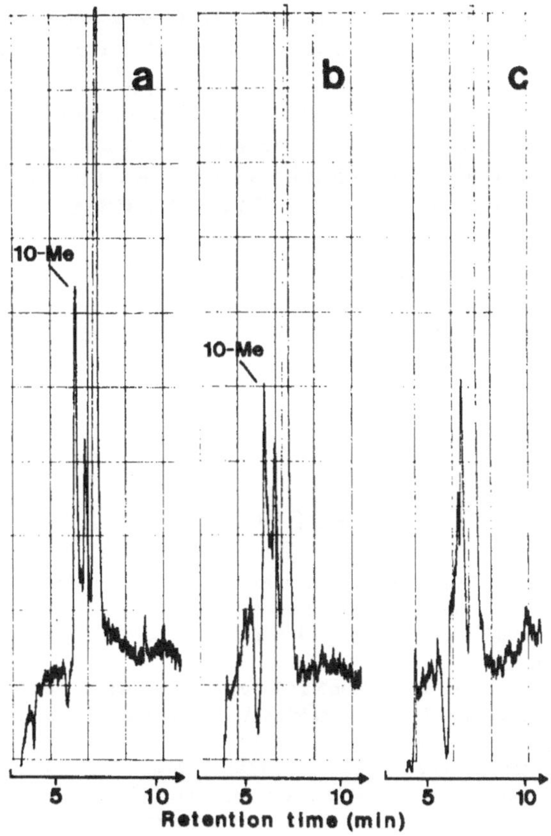

FIGURE 3. EI mass fragmentograms of sputum specimens from a patient with tuberculous pneumonia (a) and nontuberculous pneumonia supplemented (b) and not supplemented (c) with 50 pg of authentic tuberculostearic acid, using splitless injection and a 25 m capillary column. (From Larsson, L., Mårdh, P.-A., Odham, G., and Westerdahl, G., *J. Chromatogr. Biomed. Appl.*, 182, 402, 1980. With permission.)

established analytical tool in the routine identification of anaerobic bacteria by analysis of short-chain fatty acids, and of slow-growing, nonfermentative bacteria, and certain other bacteria where cellular fatty acids are studied. In addition, improved GC methods for several other microbial constituents and products such as amino acids, carbohydrates, and amines, have been elaborated and are beginning to be applied in microbiological studies. A single FSOT column enables excellent separation not only of bacterial fatty acid esters but of more polar molecules such as amino acid and carbohydrate derivatives as well. GC/MS can be used for definite structural identification of compounds separated by GC. Modern GC/MS instruments have facilities for automatic sample injection, a variety of preprogrammed analytical conditions and computerized evaluation of the resulting data where chromatograms and mass spectra are compared with reference data stored in a computer memory. Computerization improves the cost-effectiveness of GC/MS methods, particularly where a large number of analyses are to be performed. There exists comprehensive knowledge of the composition and (for anaerobes) production of bacterial fatty acids, and clearly it would be possible to construct computer programs for the evaluation of results from such studies by expressing them as identities of the microorganisms studied, with probabilities for correct identifications indicated. Detection of trace quantities of microbial constituents in clinical samples by SIM analysis is a valuable and rapid diagnostic method for, e.g., mycobacterial infections, but at present requires sophisticated analytical procedures necessitating continuous, personal attention. The use of MSD instruments for this purpose would greatly simplify such studies, provided these instruments are further developed to include facilities for CI and NI MS studies.

ACKNOWLEDGMENTS

Grants from the Swedish National Association Against Heart and Chest Diseases are gratefully acknowledged.

REFERENCES

1. Waller, G. R. and Dermer, O. C., Eds., *Biomedical Applications of Mass Spectrometry. First Supplementary Volume,* Wiley Interscience, New York, 1980.
2. Jennings, W. G., *Comparisons of Fused Silica and Other Glass Columns in Gas Chromatography,* Hutig Verlag, Heidelberg, 1981.
3. Grob, K. and Grob, J., Jr., Splitless injection and the solvent effect, in *Recent Advances in Capllary Gas Chromatography,* Bertsch, W., Jennings, W. G., and Kaiser, R. E., Eds., Hutig Verlag, Heidelberg, 1981, 455.
4. Dougherty, R. C., Dalton, J., and Biros, F. J., Negative chemical ionization mass spectra of polycyclic chlorinated insecticides, *Org. Mass Spectrom.,* 6, 1171, 1972.
5. deLeenheer, A. D. and Cruyl, A. A., Quantitative mass spectrometry, in *Biochemical Applications of Mass Spectrometry, First Supplementary Volume,* Waller, G. R. and Dermer, O. C., Eds., Wiley Interscience, New York, 1980, 1169.
6. Guerrant, G. O., Lambert, M. S., and Moss, C. W., Identification of diaminopimelic acid in the Legionnaires' disease bacterium, *J. Clin. Microbiol.,* 10, 815, 1979.
7. Larsson, L. and Mårdh, P.-A., Gas chromatographic characterization of mycobacteria: analysis of fatty acids and trifluoroacetylated whole-cell methanolysates, *J. Clin. Microbiol.,* 3, 81, 1976.
8. Larsson, L., Bergman, R., and Mårdh, P.-A., Gas chromatographic characterization of porcine and human strains belonging to the *Mycobacterium avium-intracellulare* complex, *Acta Pathol. Microbiol. Scand. Sect. B,* 87, 205, 1979.
9. Jantzen, E., Bryn, K., Hagen, N., Bergan, T., Bövre, K., Fatty acids and monosaccharides of *Neisseriaceae* in relation to established taxonomy, *NIPH Annals Oslo,* 1, 59, 1978.
10. Bryn, K. and Jantzen, E., Analysis of lipopolysaccharides by methanolysis, trifluoroacetylation, and gas chromatography on a fused-silica capillary column, *J. Chromatogr.,* 240, 405, 1982.
11. Pritchard, D. G., Coligan, J. E., Speed, S. E., and Gray, B. M., Carbohydrate fingerprints of streptococcal cells, *J. Clin. Microbiol.,* 13, 89, 1981.
12. Bengtsson, G. and Odham, G., A micromethod for the analysis of free amino-acids by gas chromatography and its application to biological system, *Anal. Biochem.,* 92, 426, 1979.
13. Fox, A., Lau, P. Y., Brown, A., Morgan, S. L., Zhu, Z.-T., Lema, M., and Walla, M. D., Capillary gas chromatographic analysis of carbohydrates of *Legionella pneumophila* and other members of the family *Legionellaceae, J. Clin. Microbiol.,* 19, 326, 1984.
14. Lambert, M. A. and Moss, C. W., Comparison of the effects of acid and base hydrolyses on hydroxy and cyclopropane fatty acids in bacteria, *J. Clin. Microbiol.,* 18. 1370, 1983.
15. Alvin, C., Larsson, L., Magnusson, M., Mårdh, P.-A., Odham, G., and Westerdahl, G., Determination of fatty acids and carbohydrate monomers in microorganisms by means of glass capillary gas chromatography: analysis of *Mycobacterium gordonae* and *Mycobacterium scrofulaceum, J. Gen. Microbiol.,* 129, 401, 1983.
16. Moss, C. W., Dees, S. B., and Guerrant, G. O., Gas-liquid chromatography of bacterial fatty acids with a fused-silica capillary column, *J. Clin. Microbiol.,* 12, 127, 1980.
17. Larsson, L. and Odham, G., Injection principles in capillary gas chromatographic analysis of bacterial fatty acids, *J. Microbiol. Methods,* 3, 77, 1984.

18. Miller, L. T., Single derivatization method for routine analysis of bacterial whole-cell fatty acid methyl esters, including hydroxy acids, *J. Clin. Microbiol.*, 16, 584, 1982.
19. Larsson, L., Acidic methanolysis v. alkaline saponification in gas chromatographic characterization of mycobacteria: differentiation between *Mycobacterium avium-intercellulare* and *Mycobacterium gastri*, *Acta Pathol. Microbiol. Immunol. Scand. Sect. B*, 91, 235, 1983.
20. Moss, C. W., New methodology for identification of nonfermenters: gas-liquid chromatographic chemotaxonomy, in *Glucose Nonfermenting Gram-Negative Bacteria in Clinical Microbiology*, Gilardi, G. L., Ed., CRC Press, Boca Raton, Fla, 1978, 171.
21. Moss, C. W., Bibb, W. F., Karr, D. E., Guerrant, G. O., and Lambert, M. A., Cellular fatty acid composition and ubiquinone content of *Legionella feeleii* sp.nov., *J. Clin. Microbiol.*, 18, 917, 1983.
22. Jantzen, E., Bryn, K., Bergan, T., and Bövre, K., Gas chromatography of bacterial whole cell methanolysates. V. Fatty acid composition of neisseriae and moraxellae, *Acta Pathol. Microbiol. Scand. Sect. B*, 82, 767, 1974.
23. Tisdall, P. A., De Young, D. R., Roberts, G. D., and Anhalt, J. P., Identification of clinical isolates of mycobacteria by gas-liquid chromatography alone: a 10-month follow-up study, *J. Clin. Microbiol.*, 16, 400, 1982.
24. Larsson, L., Jantzen, E., and Johnsson, J., Gas chromatographic fatty acid profiles for characterisation of mycobacteria: an interlaboratory methodological evaluation, *Eur. J. Clin. Microbiol.*, 4, 483, 1985.
25. Larsson, L., Valero-Guillen, P., Martin-Luengo, F., and Pacheco, F., Influence of culture medium and incubation time on the fatty acid compositions of some rapid-growing mycobacteria as analyzed by packed and capillary column gas chromatography, *Acta Pathol. Microbiol. Immunol. Scand. Sect.B*, 93, 353, 1985.
26. Holdeman, L. V., Cato, E. P., and Moore, W. E. C., *Anaerobe Laboratory Manual*, 4th ed., Virginia Polytechnic Institute and S.U. Anaerobe Laboratory, Blacksburg, 1977.
27. Larsson, L. and Holst, E., Feasibility of automated head-space gas-chromatography in identification of anaerobic bacteria, *Acta Pathol. Microbiol. Scand. Sect. B*, 90, 125, 1982.
28. Larsson, L., Holst, E., Gemmell, C. G., and Mårdh, P.-A., Characterization of *Clostridium difficile* and its differentiation from *Clostridium sporogenes* by automatic head-space gas chromatography, *Scand. J. Infect. Dis.*, Suppl. 22, 37, 1980.
29. Larsson, L., Mårdh, P.-A., and Odham, G., Analysis of amines and other bacterial products by head-space gas chromatography, *Acta Pathol. Microbiol. Scand. Sect. B*, 86, 207, 1978.
30. Moss, C. W. and Nunez-Montiel, O. L., Analysis of short-chain acids from bacteria by gas-liquid chromatography with a fused-silica capillary column, *J. Clin. Microbiol.*, 15, 308, 1982.
31. Cox, C. D. and Parker, J., Use of 2-aminoacetophenone production in identification of *Pseudomonas aeruginosa*, *J. Clin. Microbiol.*, 9, 479, 1979.
32. Phillips, K. D. and Rogers, P. A., Rapid detection and presumptive identification of *Clostridium difficile* by p-cresol production on a selective medium, *J. Clin. Pathol.*, 34, 642, 1981.
33. Alley, C. C., Brooks, J. B., and Kellogg, D. S., Electron capture gas-liquid chromatographic mass spectral identification of acids produced by *Neisseria meningitidis* in a defined medium, *J. Clin. Microbiol.*, 9, 97, 1979.
34. Larsson, L., Mårdh, P.-A., and Odham, G., Analysis of glucose catabolites using head-space gas chromatography for differentiation of some selected gram-negative species, *Curr. Microbiol.*, 4, 143, 1980.
35. Larsson, L., Mårdh, P.-A., Odham, G., and Carlsson, M.-L., Diagnosis of bacteremia by automated head-space gas chromatography, *J. Clin. Pathol.*, 35, 715, 1982.
36. Gray, J., Perks, W. H., and Birch, J., Direct gas-liquid chromatography of clinical material as an aid in the diagnosis of anaerobic lung infections, *Curr. Chemother. Infect. Dis.*, 1, 540, 1980.
37. Watt, B., Geddes, P. A., Greenan, O. A., Napier, S. K., and Mitchell, A., Gas-liquid chromatography in the diagnosis of anaerobic infections: a three year experience, *J. Clin. Pathol.*, 35, 709, 1982.
38. Brook, I., The importance of lactic acid levels in body fluids in the detection of bacterial infections, *Rev. Infect. Dis.*, 3, 470, 1981.
39. Borenstein, D. G., Gibbs, C. A., and Jacobs, R. P., Gas-liquid chromatographic analysis of synovial fluid, *Arthritis Rheum.*, 25, 947, 1982.
40. Levett, P. N., Detection of *Clostridium difficile* in faeces by direct gas liquid chromatography, *J. Clin. Pathol.*, 37, 117, 1984.
41. Pepersack, F., Labbe, M., Nonhoff, C., and Schoutens, E., Use of gas-liquid chromatography as a screenig test for toxigenic *Clostridium difficile* in diarrhoeal stools, *J. Clin. Pathol.*, 36, 1233, 1983.
42. Brooks, J. B., Nunez-Montiel, O. L., Basta, M. T., and Hierholzer, J. C., Studies on stools from pseudomembranous colitis, rotaviral, and other diarrheal syndromes by frequency-pulsed electron capture gas-liquid chromatography, *J. Clin. Microbiol.*, 20, 549, 1984.

43. Karam, G. H. Elliott, A. M., Polt, S., and Cobbs, C. G., Elevated serum D-arabinitol levels in patients with sarcoidosis, *J. Clin. Microbiol.*, 19, 26, 1984.
44. Roboz, J., Suzuki, R., and Holland, J. F., Quantification of arabinitol in serum by selected ion monitoring as a diagnostic technique in invasive candidiasis, *J. Clin. Microbiol.*, 12, 594, 1980.
45. Fox, A., Schwab, J. H., and Cochran, T., Muramic acid detection in mammalian tissues by gas-liquid chromatography-mass spectrometry, *Infect. Immun.*, 29, 526, 1980.
46. Pritchard, D. G., Settine, R. L., and Bennett, J. C., Sensitive mass spectrometric procedure for the detection of bacterial cell wall components in rheumatoid joints, *Arthritis Rheum.*, 23, 608, 1980.
47. Tunlid, A. and Odham, G., Diastereoisomeric determination of R-alanine in bacteria using capillary gas chromatography and positive/negative ion mass spectrometry, *Biomed. Mass Spectrom.*, 11, 428, 1984.
48. Maitra, S. K., Scholtz, M. C., Yoshikawa, T. T., and Guze, L. B., Determination of lipid A and endotoxin in serum by mass spectrometry, *Proc. Natl. Acad. Sci. U.S.A.*, 75, 3393, 1978.
49. Larsson, L., Mårdh, P.-A., Odham, G., and Westerdahl, G., Detection of tuberculostearic acid in biological specimens by means of glass capillary gas chromatography — electron and chemical ionization mass spectrometry, utilizing selected ion monitoring, *J. Chromatogr. Biomed. Appl.*, 182, 402, 1980.
50. Larsson, L., Mårdh, P.-A., Odham, G., and Westerdahl, G., Use of selected ion monitoring for detection of tuberculostearic and C 32 mycocerosic acid in mycobacteria and in five-day-old cultures of sputum specimens from patients with pulmonary tuberculosis, *Acta Pathol. Microbiol. Scand. Sect. B*, 89, 245, 1981.
51. Odham, G., Larsson, L., and Mårdh, P.-A., Demonstration of tuberculostearic acid in sputum from patients with pulmonary tuberculosis by selected ion monitoring, *J. Clin. Invest.*, 63, 813, 1979.
52. Mårdh, P.-A., Larsson, L., Odham, G., and Engbaek, H. C., Tuberculostearic acid as a diagnostic marker in tuberculous meningitis, *Lancet*, 8320, 367, 1983.
53. Brooks, J. B., Gas-liquid chromatography as an aid in rapid diagnosis by selective detection of chemical changes in body fluids, in *The Direct Detection of Microorganisms in Clinical Samples*, Coorod, J. D., Kunz, J. J., and Ferrano, M. J., Eds., Academic Press, New York, 1983, 313.
54. Brooks, J. B., McDade, J. E., and Alley, C. C., Rapid differentiation of Rocky Mountain spotted fever from chickenpox, measles, and enterovirus infections and bacterial meningitis by frequency-pulsed electron capture gas-liquid chromatographic analysis of sera, *J. Clin. Microbiol.*, 14, 165, 1981.

Use of Computers to Analyze, Store, and Report Results of Instrument Procedures

Chapter 13

USE OF LABORATORY COMPUTER SYSTEMS TO FACILITATE REPORTING OF INSTRUMENT-GENERATED MICROBIOLOGY RESULTS

James H. Jorgensen

TABLE OF CONTENTS

I. INTRODUCTION

Clinical microbiology is in the midst of an era of emphasis on rapid test performance and reporting, often through use of automated instrumentation. Diagnostic information regarding infectious diseases which previously took one or more days to generate may now be available in a few hours. However, the rapid transmission of important test results to physician-users has often proved difficult and at times has frustrated the effective application of rapid methods.[1,2] Ideally, clinicians should have access to microbiology data on infected patients as soon as the technical aspects of the laboratory test are completed. Results which are generated very rapidly using an automated instrument, but which are not available in the patient's medical record until the following day cannot improve clinical decision making.[3,4]

Microbiologists have long strived to provide results of extreme clinical significance (positive cerebrospinal fluid smears, positive blood cultures, etc.) by telephone communication with the patient's physician. However, such efforts are labor intensive and provide the potential for miscommunication of important details. Most laboratories also produce lengthy written preliminary reports of cultures and other examinations as an early communication in a process which may proceed to completion over a period of days or weeks in stepwise fashion.

These tedious means of communication occur at a time when the use of computers for a variety of business and personal tasks has reached an unprecedented level. Logically, computers provide the most likely means for closing the gap between improved microbiology technology and physician awareness of test results. A number of clinical microbiology laboratories currently employ computers ranging in size from small personal or microcomputers which are packaged as instrument-based "data management" systems (see Chapter 6), to large laboratory-wide main frame systems which may include on-line interfacing of certain automated microbiology instruments.[5,6]

Computerization in microbiology has not taken place easily or quickly. Unlike computerized reporting in the chemistry and hematology sections of a clinical laboratory, which can be described as the capture of numeric data defined by an expected range of numbers, computerized reporting in microbiology is complicated by the semiquantitative and often descriptive nature of microbiology test results. A patient's report may often contain a detailed description of the type of specimen and its method of collection, the quantitation and names of several microorganisms as well as their quantitative (MIC) susceptibility to a dozen or more antimicrobial agents with attendant interpretations of each MIC value. Therefore, greater effort has been required to provide the computer capacity and flexibility necessary for accurate description of various specimen types, to allow multiple sequential reports of differing lengths, and to store all of these data for later retrieval.

II. GENERAL FEATURES NEEDED IN A MICROBIOLOGY COMPUTER SYSTEM

During this period of emphasis on rapid methods and automation, all three general steps of a laboratory test must be reviewed. The initial phase involves the ordering, collection, identification, and transportation to the laboratory of a specimen. Phase 2 consists of intralaboratory performance of the desired analyses. Phase 3 is the communication of test results to appropriate physician users. Computerization has the potential to expedite and simplify each of these steps. The major emphasis by clinical microbiologists (as exemplified by this book) has until now been focused only on phase 2. However, phase 1 is vital to the efficient identification and handling of specimens, and phase 3 may be the most critical to the actual application of rapid test results by physicians.[2,4] The following should be considered as major goals for contemporary microbiology computer systems.

1. Provide the means for simple and accurate identification of patients' specimens and clear communication of test requests. Software should indicate duplicate or redundant test requests for optimal laboratory utilization and cost control purposes in the era of prospective payment for medical care.
2. Possess rapid, error-free means for entry of microbiology test results into the computer either from manual or automated instrument procedures.
3. Possess software logic to eliminate common clerical errors and to "flag" and assemble unusual test findings for review by supervisory personnel.
4. Allow prompt on-line inquiry of current or previous patient test results by laboratory personnel or physicians. Inquiry routines should be simple and "user friendly" for maximal efficiency.
5. Provide timely, easily interpreted ward and patient chart copy computer-generated reports. Reports should be easily understood by novice as well as experienced clinician-users.
6. Provide means for highlighting results of extreme clinical significance. The fact that a patient has a highly communicable pathogen should not be lost among a series of comments on the presence of usual flora or the need for better specimen collection.
7. Possibly provide a computer link with pharmacy data to alert physicians of antibiotic therapy which is inconsistent with in vitro susceptibility test results, or when less expensive drugs might be used.
8. Provide routine antibiotic susceptibility trend analyses for use by the hospital's pharmacy-therapeutics and infection control committees.
9. Provide timely assistance with hospital infection control and epidemiology efforts.
10. Assist in laboratory management efforts by:

a. Providing long-term storage and timely retrieval of microbiology data for medical, scientific, or epidemiologic purposes
b. Identifying tests which have not been completed within a normal time period
c. Facilitating the routine materials management and quality-control activities of the laboratory
d. Providing patient accounting and billing information

III. SPECIFIC FEATURES OF EXISTING SYSTEMS

A number of computer systems have been described which accomplish many of the goals on the above list. Most of these systems have been developed using custom programming, are often operated on a time-sharing basis using a relatively large computer, and have been developed to satisfy the needs of a particular institution. A few of these have subsequently been marketed as vendor-packaged systems.[6,7]

Several previous microbiology systems have utilized either computer punch cards with number codes for data entry[8-10] or special documents and optical mark sense readers.[11-13] Use of code numbers to signify the names of organisms or drugs and for descriptive comments has the disadvantage of potential clerical errors through mistranslation of numbers and presents greater difficulty for laboratory personnel in learning all possible codes. Optical document readers are a very rapid means for data entry, although the possibility exists for creation of errors through incomplete or stray markings on data entry documents.

The use of cathode ray tube (CRT) and keyboard entry has been the most common means for specimen and data entry in previously described microbiology system.[6,7,14,15] Entry has been accomplished either by typing in brief mnemonics, code numbers, or selecting from menus of possible responses using a cursor key. One of these systems[6] has provided for extremely efficient data entry through use of special function keys on a customized CRT keyboard, so that common organisms and comments can be entered using only a single keystroke. An additional advantage of keyboard data entry is that use of free text for specimen description or unusual results can be more easily accommodated.[16]

The major steps of manual data entry and data verification are potential bottlenecks which can delay the information transfer out of the laboratory. Delays may arise from limitations of the computer design itself or from the requirements of data entry by only a few clerical workers. In fact, data entry is generally the slowest and most tedious aspect of any laboratory computer system. Important design features of a microbiology system include the mode of entry, i.e. and whether it will be accomplished by a technologist or a clerk. While mark sense cards or sheets can be designed in such a way that they can be used by technologists without interrupting work flow,[15] keyboard entry by technologists offers the potential for direct interaction with the computer as soon as tests are completed.[6] However, the time required for data entry has often been judged to be too time consuming for performance by technologists, and thus has often been assigned to clerks.[10,14,16] This requires recording, collection, and transmission of some kind of report sheet from technologist to clerk which introduces redundant operations and provides the opportunity for additional errors.

Certain systems have attempted to simplify the keyboard/CRT approach through the use of formatted screens[7,14] or by use of customized keyboards.[6] In the latter example, data entry was simplified to the extent that it could be done directly by the technologists using CRT work stations in no more time than would be required to complete a form to be transferred to a clerk. The additional efficiency was accomplished in large part by programming special function keys for the most common organisms and comments. Another approach to provide time savings in results entry has been the reporting of tests as negative, unless an entry is made in the computer to the contrary.[16,17]

Some form of verification of test results is necessary to prevent the release from the laboratory of inaccurate or incomplete information. Primary verification may best be accomplished by the technologist immediately following entry of data into the computer. The second level of verification is that of routine or periodic review by the supervisor or laboratory director. However, the requirement for secondary verification of all results will likely cause undue delay in the release of results. In this regard, sophisticated computer software can be extremely useful to screen results for unusual or atypical findings and assemble them for specific review by supervisory personnel.[6,14] Use of on-line edit program can also effectively prevent entry of erroneous identification or antibiotic susceptibility data by indicating a likely error at the time of data entry and requiring that special effort be exerted to enter the atypical results.[14,18] Unusual findings which pass technologists' scrutiny can then easily be reviewed at the secondary level for final arbitration of accuracy.

IV. COMPUTER-INTERFACING AUTOMATED INSTRUMENTS

Results generated by automated instruments offer the greatest potential for rapid result transmission by directly interfacing the microcomputer which controls the instrument's function with the microbiology host computer. By placing a microbiology instrument "on-line", both the entry of test results and their final placement in a patient's file can occur without necessity of manual transcription.[6] Unfortunately, few interfaces have been developed for microbiology instruments thus far. In order to devise such an interface, it is necessary to link the internal accessioning system of the instrument with that of the microbiology host computer so that the instrument's test results can be matched with the proper patient's specimen.[18] This has been accomplished in at least one laboratory with a current microbiology instrument, the Vitek® AMS.[6] In this example, the culture accession number could be marked directly in machine-readable fields of the AMS cards, thus avoiding most keyboard transcription effort. Once the accessioning system has been devised and the instrument is fitted with a standard computer interface board (RS-232C), data can be transmitted directly to another computer. However, the host computer must in turn be programmed to decipher the instrument results accurately. This latter requirement necessitates the services of an experienced computer programmer.[18] Use of error control programs, as described above, are also crucial to expenditious information flow from an on-line instrument. These edits must, in fact, be even more comprehensive than for manually entered results if the potential for instantaneous transfer of information is to be utilized fully.

Because of the large number of available laboratory computer systems as well as the diversity of instrument types currently available, few examples of practical application of this efficient concept can be cited. Increased cooperative efforts by both instrument and computer system manufacturers will be required for the potential of on-line microbiology instruments to be realized. However, provision of such interfaces is now expected by laboratories for common chemistry and hematology analyzers. Thus, increased microbiology laboratory consumer pressure will be required to encourage the availability of microbiology instrument interfaces.

V. TIMING AND AVAILABILITY OF COMPUTER REPORTS

Computer generation of reports for use by those outside the laboratory is the final and often most visible step in use of computers in clinical microbiology. The time for generation of these reports and the method of their distribution is extremely crucial to the success of any computer system. Laboratories must attempt to provide both preliminary and final cumulative printed reports on a schedule designed to meet the practice habits of the clinician users.[3,4] Ideally, this should also be supplemented with direct access of clinicians to microbiology results via CRT terminals or perhaps small printers located in wards and clinics. For this to be truly effective, the clinician has to be confident that the computer displays the most current information. If there is a significant lag between information available within the laboratory vs. that available through a remote terminal, the value of the remote peripherals will be lost.[6] If the on-line information is not current, other less efficient means of communication, such as telephone calls to the laboratory for updated information, will continue.

VI. PRODUCTION OF HOSPITAL EPIDEMIOLOGY INFORMATION

Most microbiology computer systems provide software for retrieval of prior antibiotic susceptibility data into a periodic statistical report for use by the hospital's infection control and formulary committees as well for guidance of individual physicians in selection of empiric therapy. Such reports can be designed to summarize susceptibility data for selected microorganisms for a defined period with user defined categories of hospital locations, body sites, and patients cared for by specified physicians.[18]

One system utilizes a custom program to generate a hospital epidemiology report which includes organisms and sterile body site results selected by infection control personnel and which automatically lists the names and principal diagnoses of other patients occupying an infected patient's room.[14] Such programming has the potential to save many hours of tedious effort on the part of both laboratory and epidemiology personnel.

VII. OTHER LABORATORY MANAGEMENT FUNCTIONS

In addition to computer-generated lists of all microorganisms, comments, and antimicrobial susceptibility data which are detected by an edit system (as described above), it is possible to obtain intralaboratory reports which (1) list any organisms or responses of increased significance (all positive CSF or blood cultures, isolation of unusual pathogens such as *Brucella* spp., *Yersinia* spp., etc.),[14] (2) indicate tests which have been delayed in total turnaround time and require supervisory intervention;[14,19] (3) generate worksheets to facilitate verification of results;[6,13] (4) print hard copy records (logs) for permanent retention or as backup in case of an on-line failure of the system;[14,17] and (5) produce disk or magnetic tape for direct microfiche production records for permanent storage.[14] Software may also be written for analysis and storage of routine quality control or instrument maintenance data[20] and to assist the laboratory in materials management of media and other supplies.

VIII. THE CHOICE BETWEEN CUSTOM PROGRAMMING OR PURCHASING AN EXISTING MICROBIOLOGY PROGRAM PACKAGE

The intramural design and implementation of a microbiology computer system is a complex and time-consuming undertaking which is not appealing to most laboratories. Modification of an existing commercial package system may present equal difficulties, although for different reasons. To design and implement a laboratory-wide computer system requires a large expenditure of time and money, usually by a team composed of laboratorians and computer programmers. If done carefully, the result can be highly rewarding if a system which fulfills all of the unique needs of the laboratory results from the cooperative effort. However, updates and enhancements will undoubtedly be desired in the years following implementation of the system. Prior experience has shown that completion of all aspects of a large laboratory system may be difficult.[14,15,18] Likewise system updates and development may prove difficult if the laboratory does not employ its own programmer. It is essential in designing a new system to provide a simple means for the laboratory to update organism names and comments, provide for reporting of more or different antibiotics, and provide the ability to make changes in report formats if needed.

Selection of a vendor-packaged system may be a simpler alternative for most laboratories. For those contemplating purchase of a vendor system, it is essential to visit several laboratories of similar size and sophistication to their own which use the vendor's product. Eventual satisfaction with the system will depend largely on the purchaser's impression of the vendor's standard system. If major changes in programming are deemed essential, the vendor's likely cooperative attitude must be determined in advance. Programming changes must be detailed exhaustively both the respect to their goals and eventual function. Computer vendors often state that if extensive modifications are made to their system in one installation, other customers may demand the same improvements, thereby making the retrofitting process a never-ending nuisance for the vendor. It is also true that modification of one program may effect other programs in the same system, adding to the complexity of seemingly simple changes. Custom programming may also make the owner of a vendor-packaged system ineligible for the vendor's subsequent software updates through standard maintenance agreements, for the reasons cited above. All of these factors can contribute to a lack of cooperative interaction when a laboratory wishes to customize a vendor's existing packaged system.[21] System purchasers should be extremely wary of general statements such as "the laboratory can design its own report formats." Such details must always be contingent upon restrictions inherent in the design of any particular computer system.

If an existing microbiology computer system is being replaced by a newer system, the adaptation of laboratory personnel and physician users to the new system will directly related to prior circumstances, i.e., it is often easier to computerize a laboratory which has been using manual reporting methods, than to switch from one computer system to another.[21] Inevitable comparisons will be made between two computer systems, making adjustment to a new system difficult in some instances. In order to avoid such difficulties, its is wise to assemble a group of "users" to review the merits of any contemplated new system prior to purchase or initiation of new programming.

Clinical microbiology continues to be the most demanding area of the laboratory with respect to computerization. Larger laboratories, especially those in academic medical centers, often set high standards for the accomplishments of a computer system which may not be easily met of commercially marketed computer packages. Therefore, the decision to custom design an intramural system or to buy a vendor packaged system must be carefully weighed based upon local resources, demands, and professional interests.

IX. SUMMARY

Despite some shortcomings of existing microbiology laboratory computer systems, several innovations have occurred through development of these systems, including a few instrument interfaces,[5,6] use of dedicated CRT keys,[6] automatic reporting of negative results,[16,17] and the provision of error control or edit programs.[6,9,14] All of these features should be considered for inclusion in new systems in order to provide a truly efficient and accurate means of transmitting microbiology data. The full potential of automated instruments will only be realized through sophisticated application of computer technology for final communication of test results to clinicians.

REFERENCES

1. Matsen, J. M., Rapid reporting of results — impact on patient, physician, and laboratory, in *Rapid Methods and Automation in Microbiology,* Tilton, R. C., Ed., American Society for Microbiology, Washington, D.C., 1982, 98.
2. Ellner, P. D., Rapid reporting — does it really affect patient care?, in *Rapid Methods and Automation in Microbiology,* Tilton, R. C., Ed., American Society for Microbiology, Washington, D. C., 1982, 106.
3. Matsen, J. M., How rapid should rapid methods be?, *Clin. Microbiol. Newlsett.,* 45, 164, 1982.
4. Matsen, J. M., Means to facilitate physician acceptance and use of rapid test results, *Diagn. Microbiol. Infect. Dis.,* in press.
5. Kozub, W. R. and Morrow, D. J., Autobac in a computerized microbiology laboratory, in *Rapid Methods and Automation in Microbiology,* Tilton, R. C., Ed., American Society for Microbiology, Washington, D. C., 1982, 28.
6. Ryan, K. J., Methods for rapid transmission of microbiology reports, *Diagn. Microbiol. Infect. Dis.,* 3, 335—385, 1985.

7. Gavan, T. L., Brotzman, B., and Smith, C., Description of a clinical microbiology information system, in *Rapid Methods and Automation in Microbiology,* Tilton, R. C., Ed., American Society for Microbiology, Washington, D.C., 1982, 19.
8. Vermeulen, G. D., Schwab, S. V., Young, V. M., and Hsieh, R. K., A computerized system for clinical microbiology, *Am. J. Clin. Pathol.,* 57, 413, 1972.
9. Farrar, J. L., Brogan, T. D., Moulding, T., Taylor, R., and Page, M., Use of a computer for producing microbiological reports and for data storage and processing, *J. Clin. Pathol.,* 28, 580, 1975.
10. Spraberry, M. N., Computer processing of microbiology data — part of a total laboratory system, *Am. J. Med. Tech.,* 35, 77, 1969.
11. Ayliffe, P. F. and Chalke, R., A computer system to perform record handling and reporting for a hospital microbiology laboratory, *Med. Lab. Tech.,* 30, 363, 1973.
12. Bergqvist, F. and Bengtsson, S., The BACTLAB system — a data system for bacteriological routine, *Comput. Programs Biomed.,* 4, 144, 1975.
13. Andrews, H. J., and Vickers, M., An assessment of one year of computer-assisted microbiology reporting at Charing Cross Hospital, *J. Clin. Pathol.,* 27, 185, 1974.
14. Jorgensen, J. H., Holmes, D., Williams, W. L., and Harris, J. L., Computerization of a hospital clinical microbiology laboratory, *Am. J. Clin. Pathol.,* 69, 605, 1978.
15. Lowrie, D. L., Elin, R. J., Gill, V. J., Lewis, T. L., MacLowry, J. D., and Witebsky, F. G., Microbiology subsystem of a total, dedicated laboratory computer system, *J. Clin. Microbiol.,* 10, 861, 1979.
16. Kunz, L. J., Poitras, W. J., Kissling, W. J., Mercier, B. A., Cameron, M., Lazarus, C., Moellering, R. C., and Barret, O., The role of the computer in microbiology, in *Modern Methods in Medical Microbiology,* Prier, J., Bartola, J., and Friedman, H., Eds., Baltimore, Md., University Park Press, 1975, 181.
17. Bhargava, B., Design and implementation of noncoded computerized microbiology system, *Lab. Med.,* 15, 334, 1984.
18. MacLowry, J. D. and Ryan, K. J., Computers in clinical microbiology, in *Manual of Clinical Microbiology,* 4th ed., Lennette, E. H., Balows, A., Hausler, W. J., Jr., and Shadomy, H. J., Eds., American Society for Microbiology, Washington, D.C., 1985, chap. 7.
19. Pribor, H. C. and Scarry R. L., A microbiology computer system, *Lab. Management,* October, p. 40, 1984.
20. Buck, G. E. and Balfour, L., Microcomputer storage and analysis of quality control data in clinical microbiology, *Lab. Med.,* 16, 359, 1985.
21. Jorgensen, J. H., The choice between custom programming or purchasing a vendor packaged system for computerization of a clinical microbiology laboratory, in *The Role of Clinical Microbiology in Cost-Effective Health Care,* Smith, J. W., Ed., College of American Pathologists, Skokie, Ill, 1985, part 13, chap. 4.

Chapter 14

THE ROLE OF MICROCOMPUTERS FOR DATA ANALYSIS AND STORAGE

George E. Buck

TABLE OF CONTENTS

I. INTRODUCTION

Computers have contributed significantly to effective storage and analysis of information in business and industry during the last 20 to 30 years. Significant progress has also been made in this regard in the clinical laboratory, although perhaps at a somewhat slower rate. The recent introduction of low-cost microcomputers has added an entirely new perspective to the field, however. With this, the cost of computer equipment has been reduced to the point that almost any laboratory can have access to a computer and, indeed, many laboratories already do. A number of systems are currently available for clinical microbiology laboratories in which a microcomputer is utilized to acquire data from a particular instrument to record and summarize results. The intent of this chapter is not to present an exhaustive review of these systems, but to explore some other uses recently developed for microcomputers in clinical microbiology. Since computer systems are so common now, many people have a basic understanding of their function; for those who do not, it may be beneficial to discuss several basic concepts and terms before proceeding further.

II. BASIC CONCEPTS AND TERMINOLOGY

Science fiction movies tend to create the image of a computer as a mysterious box with innate powers. In truth, a computer is really just a large bank of memory slots where numbers or characters are stored. A series of instructions (the computer program) controls the manipulation of the data. The person who writes the program determines what is done with the data, thus the computer can really do nothing on its own. In any computer there is a central component, called the central processing unit (CPU), which performs these basic actions. Figure 1 illustrates the main parts of the CPU. Main memory is a series of locations which can be thought of as a collection of mail slots. Each slot is numbered to identify it uniquely and each slot can contain one item of data (a number or character). Generally the number of memory locations is quite large, in the thousands or hundreds of thousands. Part of the memory holds program instructions and the remainder contains the data to be manipulated. The control unit is the circuitry for interpreting the program instructions and routing data to the various units of the computer. The arithmetic and logic unit (ALU) is the unit that performs arithmetic function, such as addition, multiplication, or division, etc., and logical comparisons of numbers. The registers are small memories that hold temporary results of the data manipulations.

During actual operation, the control unit begins with the first step in the program. After the first step is executed the control unit moves to the next instruction and continues on in this way until the end of the program. The program steps consist of instructions to fetch data from memory, perform a comparison or arithmetical procedure, and then store the results of the manipulation back in memory. This entire process continues as a cycle; data items are fetched from memory, some function is performed on the data, then the results are stored back again. The speed of the electronic circuits allows this cycle of events to occur thousands of times each second. This accounts for the amazing speed of the computer. Thousands of calculations can be done or thousands of names can be stored or compared in a second.

Each data item or program instruction in memory is represented by binary number, which is a string of zeros or ones. The number of digits used by any particular computer is fixed, but is different in different types of computers. Some computers use binary numbers with 8 digits, others 16 or 32. In computer terminology each digit is called a bit (from binary digit) and, by convention, 8 bits are called a byte; collections of bits larger than 8 are called a word.

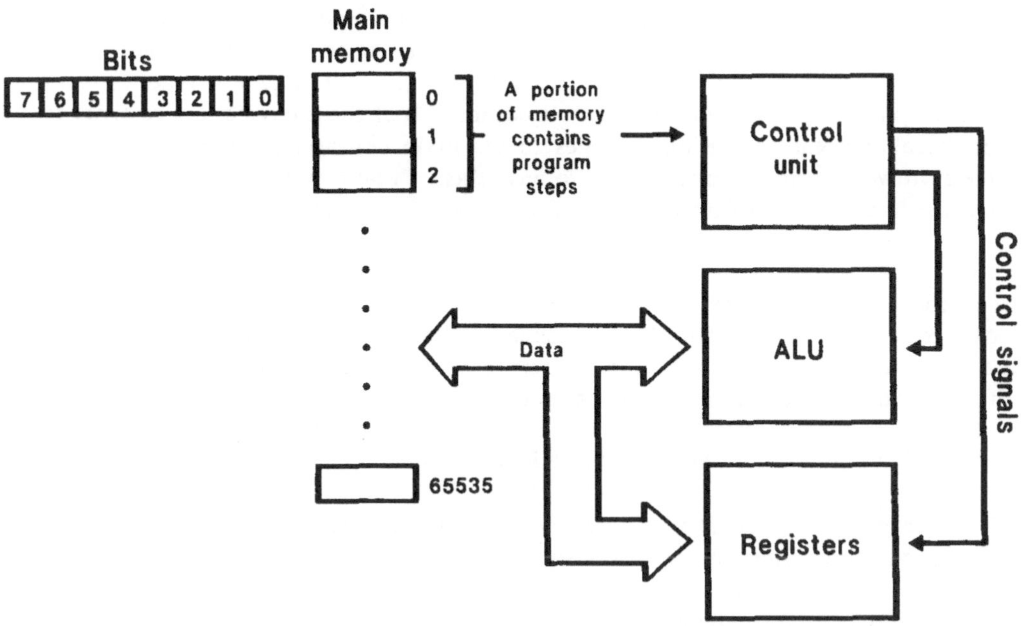

FIGURE 1. Major components of a computer.

This difference in the size of binary numbers used in operation has far-reaching consequences in regard to the characteristics of a computer. Each memory location is given a unique number to identify it when items of data are read to or from it. Therefore, computers that use a small number of bits are more limited in the amount of memory, and thus in the size of programs and the amount of data, then computers that use a larger number of bits for calculations. Memory size is usually described in increments of 1024 (2^{10}), but rounded off to 1000. Each segment is referred to by the letter K (for kilo). This 1K of memory is 1024 memory locations, 2K is 2048, 4K is actually 4096, 16K is actually 16,384, 32K is 32,768, and 64K stands for 65,536. Since the memory size of a computer is an important characteristic in describing it, these terms are used often by computer users.

Another factor in the operation of the CPU that has a large effect on the operation characteristics is the cycle time. As already mentioned, the CPU operates by copying the value in a particular memory location to one of the registers (typically, computers will have several). For example, if a number that needs to be added to another number is in memory location 250 (in binary this is 11111010), the control unit goes to location 250, transfers a copy of the value located there to one of the registers, then finds the other number to add to it, completes the actual addition operation in the register, then transfers the answer to a third memory location. All of this is directed by the program. Although the electronic processes happen very rapidly, they do take some time, and the speed of this operation determines how fast a computer can function. Typically the time necessary for an operation will range from a few milliseconds to microseconds. Although both of these are far faster then a human can function, this means there can be as much as a tenfold difference in speed of operation.

The CPU is the heart of the computer, but other hardware items are necessary for complete function. There needs to be a way for information to get into the computer and transmitted back to the human environment. Although a number of different devices have been developed for this, at the present time the most common device for transmitting data to the computer is a keyboard and the usual means for data output is a video screen or printer. In addition, since the main memory is limited and is nonfunctional when the power is turned off, there needs to be a means of permanent storage for large amounts of data. Generally, this is done with magnetic tape, or disks of metal or plastic covered with magnetic medium. All of these devices are referred to as peripherals since they are appendages to the main component, the CPU. To be complete, however, the computer needs to have them. A large number of other peripheral devices have been developed, but these are beyond the scope of the current discussion.

This has been a brief overview of several main concepts and terms used in connection with computers. In summary, the main points to be remembered are that the capabilities of a computer are determined by the speed of the electronic transfers (cycle time) and the amount of memory, which in turn governs the size and, thus the capabilities, of the program, and the amount of data that can be conveniently accessed.

III. CAPABILITIES OF CURRENT MICROCOMPUTERS

When computers were first developed in the 1940s, the circuits for operation were made with vacuum tubes. Many thousands of vacuum tubes were used in the first generation computers, and the capabilities of such hardware were not very great, thus the early computers were expensive, large, and performed poorly by toady's standards. With the introduction of transistors, it became possible to condense the electronic circuits to a much smaller size and the performance and reliability improved significantly.

The innovation that made microcomputers possible is called large scale integration (LSI). In essence, this is the process of etching thousands of electronic circuits onto a single, tiny chip of silicon. As a result of this, the capability of many thousands of transistors can be reduced to a miniscule size.

The general configuration for microcomputers is one main chip which contains the three essential components of the CPU, the control unit, the ALU and registers; this is called a microprocessor. Generally the majority of associated memory is added separately with additional memory chips, allowing flexibility in memory size. The newer microprocessors coming out now, however, are moving in the direction of having memory integrated into the microprocessor, a true computer on a chip.

The basic difference between a microcomputer and a minicomputer or so-called mainframe is simply that the circuitry for the microcomputer CPU is on one chip as a result of LSI. Because of the technological difficulty in designing small chips to contain all of the necessary electronic apparatus, this means microcomputers have generally been limited to 8- or 16-bit processors and thus smaller amounts of memory. Prior to the introduction of the IBM®-PC several years ago, the usual configuration for microcomputers was an 8-bit microprocessor and 64K (remember, that's actually 65,536 bits) of memory. With the introduction of the IBM®-PC, with its 16-bit Intel® 8088 microprocessor, the *de facto* standard has become a 16-bit microprocessor and 128K to 512K of memory. This is enough memory for relatively sophisticated programs but still not in the minicomputer range. The other characteristic of microcomputers is that they generally are slower. Most minicomputers or main frames have 32-bit processors or larger. This allows memory sizes of 1000K (1 megabyte) or more and these processors operate extremely fast. With these characteristics they are able to have multiuser and multitasking capabilities, i.e., they allow what seems to be more than one user or task to go simultaneously. In actuality the operations are not simultaneous, merely fast enough that the available memory can be switched back and forth between users or tasks, and it appears to be simultaneous.

In summary, then, the technology of current microcomputers is such that they are relatively limited in memory and somewhat slower than minicomputers or mainframes, although much smaller. Therefore, microcomputers in use at the current time are not able to execute programs that are extremely sophisticated, because the more complex a program is the more memory it requires. In addition, microcomputers are roughly 5 times slower than larger computers. This means that searching and organizing data takes just a bit longer. Finally, because of the combination of the above characteristics, microcomputers are more suited to single users, rather than allowing several users or tasks to operate simultaneously. The one big advantage of a microcomputer, of course, is the low cost. This is because LSI manufacturing technology allows the chips to be mass produced. Even with the limitations noted above, however, there are still a number of useful applications.

IV. APPLICATIONS FOR MICROCOMPUTERS IN CLINICAL MICROBIOLOGY

The most obvious use for computers in the clinical microbiology laboratory is storage and reporting of routine laboratory results. This has several potential advantages: the data, once stored, can be arranged by the computer so that results can be presented in a neat, logical format; the data can be analyzed and compiled for ancillary reports, such as antibiotic summaries, epidemiological reports, and work-load reports, much more readily by a computer then searching through a stack of manual reports; the data can be stored in a more condensed and accessible form than in volumes of log books or data sheets.

Although microcomputers can be programmed fairly readily with easily learned computer languages such as BASIC, the skill required to write high quality software is sufficient that most users need to rely on commercial software. However, as more and more people become interested, some public domain software for reporting routine laboratory results may eventually become available. Several commercial software packages have recently been marketed or are in the process of development. Because of limitations of microcomputer hardware previously noted, the microcomputer report systems available at the present time are limited to laboratories serving hospitals with about 300 beds.

One of these systems is marketed by Health Development, Inc. (Columbus, Ohio) and is called Micro-IMS, for Microbiology Information Management System. This is a set of programs that run on the IBM®PC-XT, which has one diskette drive and an integral fixed disk, which has the capacity to store about 10,000,000 characters of data. The cost, including hardware, is about $17,000. A full laboratory system, which includes the Micro-IMS system along with programs for handling other clinical lab data, is also available, at a cost of $30,000, including hardware. The programs are written in PASCAL and are menu driven, i.e., an initial display contains a list of choices and selecting one of the choices leads to another list of choices being displayed. Both the programs and data are stored on the integral fixed disk of the XT and the diskette is for permanently storing completed results, making backups and introducing new programs to the fixed disk. The main menu contains a list of the main functions of the system:

1. Accessioning, discharge, test request
2. Result entry
3. Data archiving
4. Dictionary definition
5. Data management

6. Result inquiry
7. Utility functions
8. Password definition

When a specimen comes into the laboratory a file for the patient demographic data and specimen information is created using the first function. The patient demographic data need to be entered for any particular patient only once and it remains on file for subsequent specimens. The user then enters information about the specimen and selects the tests to be performed. Each type of test has a numerical code associated with it, and this code is entered. After a batch of specimens are entered, the computer generates a work sheet containing blanks to write in results and labels to label tubes and plates.

The second option on the main menu is result entry; the purpose for this is obvious. The results for organisms are entered by code; each organism is assigned a numeric code. There is room for five isolates to be entered on each specimen. An attractive feature of this system is the ability to enter either categorical results for sensitivity testing (S,I,R) as MIC results, or both, for any isolate.

The third selection on the main menu is for housekeeping functions. This allows making backup copies of data and archiving specimens no longer active. The fourth function is for defining the dictionaries. A number of things kept by the system are user-defined and can be changed any time: source/site codes, result codes, drugs used for susceptibility testing, test numbers, pricing, hospital departments, and client doctors. In addition, the format of the results report and work sheets can also be varied.

The data management function produces written reports on various types. One of the strong features of this system is the number of different types of reports it can generate. In addition to preliminary and final patient reports, MIC interpretive reports, showing usual doses of antimicrobial agents in relation to MICs of isolates, can be generated and a number of summary reports can also be printed, including reports showing antibiograms, infection surveillance, CAP work load, daily and monthly summaries of tests done.

The purpose of the "result inquiry" function is obvious. The utility function allows several things such as setting the time and date. The password function enables or disables the requirement for a user to enter a password to sign onto the computer, depending upon laboratory preference. Another interesting feature of the system is the ability to set the system so that the carriage return need not be used on data entry fields. When the carriage return feature is off, no carriage return is necessary and this speeds up data entry. Provision has also been made for interfacing with a hospital computer system for automatic transfer to patient data avoiding the necessity for keying in patient data.

Another microcomputer system that has just become available for microbiology was developed by LAN Craft, Norcross, Ga., and is called the MBS system. It developed from an on-line data acquisition system for use with the Micro-media® MIC test panel reader, and is called SuperComp in this form. However, it is evolving into a full laboratory system for general use. Currently it uses Televideo microcomputer equipment, although plans are in progress to implement the programs on the IBM®PC-XT and PC-AT.

One of the unique features of this system is that it utilizes a local area network, i.e., several microcomputers linked together to share programs and data. At the present time two microcomputers can be linked together. In this way several users can have "simultaneous" access (it is not really simultaneous, but sequential access with sufficient speed that it seems simultaneous) to the data and, thus, is a multi-user system.

Future plans call for development of a larger network system containing many independent processors linked together and sharing common disk storage, which would have significant potential for use in larger institutions.

This system is not menu driven, the main functions are invoked by entering a two-character mnemonic code. For example, once the system is started, to enter patient demographic data the user types in AD (for "administrative data") and a screen is displayed to enter the information. Typically when a specimen comes into the lab, the specimen entry function (AS) is initiated, then the patient number is entered and the computer checks to see if the specimen is on file. If not, the AD function is automatically called and patient data are entered. If the patient is on file, however, processing proceeds for entry of information about the specimen. Other main functions, such as printing reports, changing code definitions, and retrieving results are similarly initiated using a two-character code. A list of the other mnemonic codes is as follows:

AS — Ad/specimen entry
AD — Administrative data
BA —Batched reports — print
CD — Code definitions
FL — Formulary lists
IN — Incidence reports
MR — MIC report — print 1
AC — New specimens list
TR — Patient test results
DT — Set date/time
SP — Specimen test results
ST — Standard text
SR — Summary report — print 1

After the major programs are initiated, the results are also entered by codes, although in this case the codes are up to four characters. These codes are user defined and can be changed any time. If the user cannot remember the necessary code, a HELP screen is available at each step to provide the necessary information. To take advantage of the HELP function, just depress the HELP key and a "window" is formed on the screen, i.e., part of the current screen is erased and another display is shown. When the user is finished with the "window", the previous screen is re-formed. The "window" feature is another unique feature of this system and is also used for comments. In this case a window is opened up where the text comment can be typed in, when the comment is finished the original screen is re-formed. The use of the "window" feature allows different operations to take place on the same screen during the same period of time.

Several different types of reports can be printed with this system, including a list of specimens processed during any specified period of time, routine patient results, MIC interpretive reports, and antibiograms. In addition, there is a report generator, which allows almost any type of data stored to be printed with a format defined by the user. The system still bears the marks of having evolved from a data acquisition system, however, since it obviously is intended for an MIC system and does not allow categorical susceptibility testing results (S,I,R).

In addition to these more generalized system, more restricted data storage and reporting systems are available. One is a series of specialized programs marketed by BactBob Software (San Francisco, Calif.) for routine storage and reporting of data for syphilis, rabies, rubella and TB testing.

The programs were written for the Hewlett-Packard HP-150, which has a touch-sensitive screen, so control for the program and data entry is by touching the screen. In the syphilis serology program the basic menu allows a choice of functions: adding new patients to the system; printing a work sheet; entry of results; printing a daily log. Once the patient information is stored, a work sheet can be printed for initial recording of the data, or else the information can be directly entered into the computer, using the screen as a work sheet. In this case, the specimen number is entered, the patient data is automatically filled in on the screen, and the results are entered. All of the results are entered by touching the screen. For numerical values, a representation of a keypad is displayed as part of the screen for data entry and touching the number on the "keypad" enters the number.

The other programs are similar; they consist of a series of menus to proceed through the programs. Control of the program operation and data entry is by touch-entry on the screen. In the rabies program, the case information and results of fluorescence antibody testing are stored and printed out on a summary sheet. For rubella testing, patient information plus results of the serological test are entered and stored. For the mycobacteriology program, patient information and results of direct smears and culture are recorded, plus results of biochemical tests, if cultures are positive. In all of these cases, the computer can be queried directly or a summary sheet can be printed out. In effect, the results allow the microcomputer to be used as an automated logbook where results can be accessed several different ways.

Another area where significant progress has been made in utilizing microcomputers in microbiology is in compiling infection control data. One commercial package for this is called dNOSO, marketed by Infection Control Consultants, Inc. (Lafayette, Ind.). This package is built around a popular data base management system for microcomputers, called dBASE II, which has its own programming language, so the programs are in dBASEII, language and the data is manipulated by the integral data base management system. The program is available strictly for the IBM® PC or PC-XT. To start processing the user types DNOSO and a main menu comes on the screen allowing data entry, editing, printing reports, etc. If the data entry option is selected, a series of screens is displayed for entering the patient's name, address, hospital number, hospital location, diagnosis, admission date, and site of infection, along with whether the infection is nosocomial in origin. Microbiological information about the organism is then entered. This includes a code for the organism name and susceptibility pattern. Up to three different organisms can be tracked for one patient. All of this information is stored and compiled for reports. Reports can be generated at any time and can cover any time period desired. Both line listing, showing all patients with nosocomial infections or community-acquired infections, and summary reports can be obtained the summaries show infections by hospital service and body site. The information is stored on a diskette, and there is room for 150 to 200 patients on one diskette; however, summary tallies can be carried forward onto subsequent disks, so that summary information for an extensive period of time can be stored, an graphs can also be printed for the data. With this type of approach, the tedious calculations can be eliminated. The program also has the capability of summarizing the information in graphic form.

One area of use that is just beginning to be investigated is storage and analysis of quality control data. The author has developed some programs for use in our laboratory for this purpose. These programs have been in use for about two years now. A recent publication gives greater detail.[1]

The computer used was a CompuStar® microcomputer (Intertec Data Systems, Columbia, S.C.), with 64K (65,356 bytes) of memory, two double-sided, double-density

5¼-in. diskette drives (350K of storage per disk) and was also attached to a fixed disk drive with 10 megabytes (10,000,000 bytes) of storage. The use of the fixed disk while not essential provides large storage capacity and rapid access, thus facilitating handling of data.

A series of programs was written in BASIC to store and retrieve the data. At the present time only programs for media, reagents and susceptibility testing have been developed. In each instance there is one program for data entry and storage and another for data analysis.

An overriding consideration for all data entry was that as much background information as possible, such as pH of media, drugs tested, etc., be stored on the disk and only entered when necessary. This facilitates data entry by minimizing the amount of data actually entered during routine use, but also required that each program have facility for entering and changing background information. In addition, it was necessary to provide the ability to correct errors in data previously entered. Other features deemed necessary included the ability to recall information for any specified period of time and print full results or a summary either on the screen or a printer.

When results of routine testing of media are available the program for storing the information is invoked by typing in MEDIA, the computer then responds by asking for the date. When the date is entered an initial menu allows a choice of entering new data or modifying previously entered information. The computer then lists a prestored list of media routinely tested. The medium to be entered is selected by number and a data entry screen is printed. The lost number is entered along with data entered, expiration date, and number received. The organisms used for that medium and expected reactions are printed, then a prompt is printed to designate whether the result was acceptable or not: Y or N is entered and the cursor moves to a common field. A short comment can then be entered, if desired, or if the RETURN key is pushed, then the next organism/reaction combination is printed, and the cursor moves to the next prompt. When all of the reactions are entered, the computer asks if the medium was sterile, then up to five lines of comments can be entered. At this point pressing RETURN sorts the data, and entering X causes the entry process to be restarted. The computer then returns to the media list for another selection. Because the organisms and reactions are prestored, the entry process takes just a few seconds. If changes need to be made in data previously entered then the second option is selected on the main menu and a similar screen is displayed, but for a test previously entered. In the retrieval program a full report can be printed either on the screen or printer, or the computer will go through recorded data and summarize the results for any period of time specified.

The program for entering susceptibility data is invoked by typing in SENS. Here again a main screen comes on giving several choices. To enter new data, the first choice is selected. Information is then displayed on pH of the media, depth of plates, etc. This information is prestored and can be changed at entry time, but only needs to be changed when new lots of materials are used. The antimicrobial agents are then listed on the screen one at a time in alphabetic order along with the disk content, brand, lot number, expiration date, and zone size limits for the organisms used. A prompt (question mark) is printed at the point where the zone size should be entered. As with the process for media, the drugs tested and descriptive data for each drug are stored and only changed when necessary, thus allowing data entry to be streamlined. Data for one day's test can be entered in about 1 min. Option 2 on the main menu allows review of data previously entered, option 3 allows corrections to data if necessary, and option 4 allows addition, deletion, or changes in antibiotic information. These last features are only rarely needed. The retrieval program prints full reports for any day on the screen or printer, or a summary showing average zone sizes for all drugs with the three test organism, or else a report showing the drugs and dates tested where zone sizes were out of limits. The results of the tests are stored on the fixed disk until 3 months of data are accumulated, then the data for the quarter is transferred to a diskette and the files on the hard disk are reinitialized. The system has worked relatively well but could be improved by making data retrieval more flexible and providing graphic output. Newer data base management systems that incorporate extensive graphics capabilities have recently become available that would be ideal for this.

IV. FUTURE PROSPECTS

The above examples demonstrate current capabilities and uses of microcomputers in clinical microbiology. Virtually everyone, though, is aware that rapid advances are taking place in this field. In particular, three areas of progress are now in development which will have a large impact. First is progress in LSI to pack more circuits on smaller chips. Since the advent of the IBM® PC the *de facto* standard for microprocessors has been the 16-bit chip. Already, though, improvements have been made and IBM has just announced the IBM® PC-AT (for "advanced technology") which uses the 80286 microprocessor; this has a faster cycle time and additional ancillary circuits. In addition, 32-bit microprocessors have just been developed and will be used more and more in microcomputers in the future. The Apple Macintosh® uses the Motorola® 68,000 chips, a 32-bit chip. These more powerful machines have been referred to as "super-micros" or desk top minicomputers. In effect, the advances in hardware construction have increased the capabilities of microcomputers to the point where the distinction between microcomputers, minicomputers, and main frames is blurred. An, in fact, in the future this distinction may not be made. All ordinary computers will probably be the size of current microcomputers with some distinction based on the complexity of hardware. When this happens, microcomputers will essentially have the capabilities of any of the large systems available now.

Another hardware development that has evolved during the last few years is distributed processing. This concept is based on the idea of distributing the computing tasks over several smaller, linked computers rather than one larger machine. The hardware can thus be simpler and cheaper and each individual user has greater access to the overall system. This idea is already in use, to some extent, in hospitals where the laboratory computer is interfaced with the main hospital computer, to exchange information on patient data. It can be carried further, though, to have a hierarchy of computers linked together to exchange information. In this way, the computing tasks are divided up to be done on a local level, allowing less complex, and thus cheaper, hardware and ready access by all users. A similar concept is a local network where a group of independent microcomputers is linked together in a network to share peripheral devices such as disk storage units, printers, etc. This could provide an extremely flexible system composed of 1 to 100 or more units. A small laboratory could just have one unit, providing a cheap, entry-level system and larger laboratories could add additional units to have a multi-user system. Upgrading to a larger system would be a matter of adding additional hardware modules and possibly making slight software changes. Ultimately, computer systems may evolve into some combination of these concepts, such as the configuration shown in Figure 2, where a hospital computer passes information to the laboratory computer, which in turn connects one or more local area networks or hierchy of subordinate computers. Whichever way it ends up, networking and extensive interfacing will probably be used more and more to provide greater flexibility and accessibility.

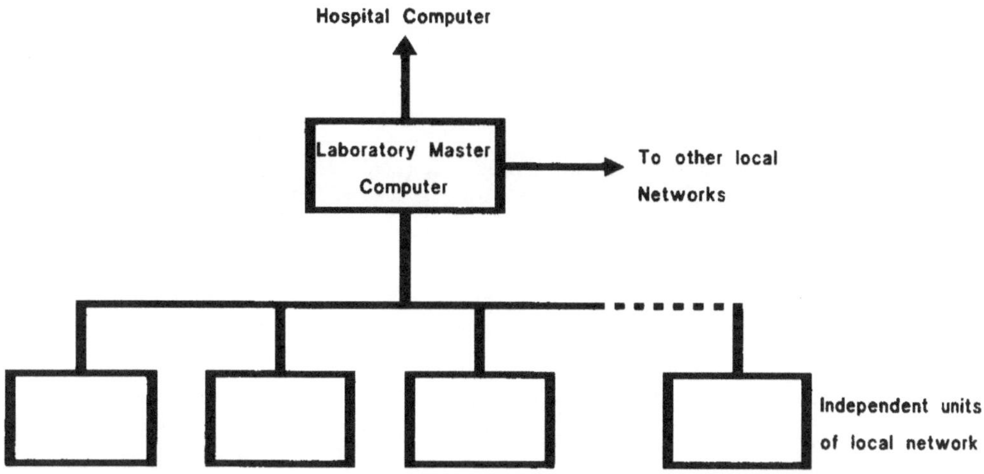

FIGURE 2. Possible hardware configuration for a laboratory computer using distributed processing with local area networks.

The other area where rapid changes are taking place is software sophistication. In the early days of computers the major cost involved in developing a computer system was hardware; the cost of programming was generally not too significant. With the recent drastic reduction in hardware, though, the relative cost of hardware has been reduced while the cost of software developed has pretty much stayed the same. To overcome this, there is extensive effort to make software easier to develop and change. Programs whose function is to write other programs, so-called "program generators", have recently been developed. In addition, the wide availability of microcomputers has created a demand for software that is easy to use and can be used on different computers. These factors, coupled with greater sophistication of hardware, has spurred development of easily used programs with extensive capabilities. This is exemplified by the recent emergence of several very powerful data base management systems which have elegant graphics capabilities and an easily learned internal programming language that allows flexibility in the format of data storage and retrieval. In the future, use of this type of software with inexpensive yet powerful hardware should permit laboratories to store almost all data electronically and summarize or compile it readily and rapidly, with elegant graphic representation, to make handling of information easier, more accurate and more meaningful.

REFERENCES

1. Buck, G. E. and Balfour, L., Microcomputer storage and analysis of quality control data in clinical microbiology, *Lab. Med.*, 16, 359, 1985.

Acceptance of Current Automated Microbiology Instruments

Chapter 15

ACCOMPLISHMENTS OF CURRENT AUTOMATED MICROBIOLOGY INSTRUMENTS

Marie Pezzlo

Automation has had a major impact on the diagnostic microbiology laboratory during the past 10 years and this trend will undoubtedly continue through the next decade. The increased interest and growing use of automation is a response to the needs of clinical laboratories for more simplified test procedures, reduction in turnaround time, the ability to perform direct testing of certain specimen types, and a real need to reduce labor costs. Automation has already had an impact on diagnosis of infectious diseases by making it possible to obtain information on the etiologic agent of an infection along with the appropriate antimicrobic needed for treatment much sooner than before. As a result, the patient's hospital stay may be shortened.[1] Both the patient and the laboratory have shared the benefits and accomplishments of automation thus far.

The wide variety of automated systems for use in the diagnostic microbiology laboratory have been discussed in previous chapters of this text. The major advantage of most of these systems is the automated measurement of a variety of physicochemical characteristics of microorganisms and their analysis by some sort of computer pattern recognition system. The degree of automation varies with the different systems. Ideally, those systems which may be referred to as "walk away" provide the greatest advantage. They require the least amount of technical time and, therefore, provide a decreased labor cost. Availability of results are not dependent upon the presence of the laboratory staff and results are often available within a few hours.

As we consider the advantages of automation with regard to patient care, we must continually be mindful of the purpose of clinical microbiology. This purpose should be to provide relevant data required for patient care as accurately and economically as possible and as quickly as is feasible, especially in cases of life-threatening infections. Certain life-threatening infections such as meningitis, septicemia, and endocarditis must be empirically treated in the early stages of the disease process before the organism identification and antimicrobial susceptibility pattern is known. However, rapid reports can have more significance with the seriously ill patient in correctly guiding continued therapy.[2] Rapid organism detection in blood cultures is necessary in this sense for optimal patient care, and rapid susceptibility results are important for successful modification of initial therapy. In order to microbiologic data to have a greater impact on patient care, results must be provided as quickly as possible, preferably within hours of specimen collection. In one institution, physicians, when polled regarding their preferences, chose same-day automated semiquantitative antimicrobial susceptibility results over more quantitative test results which would require overnight incubation.[1] Presently, the only systems providing same day antimicrobial susceptibility results are instrument methods.

In addition to patient with life-threatening infections, other patient groups may also benefit from rapid results. These include immunocompromised hosts, burn patients, patients with intravascular or urinary catheters, shunts, and prosthetic devices, and those receiving antimicrobial agents. Even ambulatory patients with a urinary tract infection may benefit from rapid results which may influence the physician's clinical decision regarding the type and duration of therapy. Automated urine screening methods which are currently available provide same-day results, some within a few minutes.

Close cooperation and communication between microbiologists and clinicians is needed to select optimal use of rapid automated tests for the betterment of patient care.

What has automation accomplished so far for the laboratory? Instruments may now be used to accomplish tasks which constitute a majority of the workload in most diagnostic microbiology laboratories. These tasks include detection and, in certain situations, quantitation of microorganisms in sterile fluids such as blood and urine; prompt identification of a variety of etiologic agents; semiquantitative antimicrobial susceptibility testing; assay of antimicrobial agents in body fluids, and direct detection of microbial antigens in body fluids, i.e., cerebrospinal fluid. The parameters, accuracy, reduced turnaround time, efficiency, and cost-effectiveness have been well addressed by currently available instruments.

Diagnostic microbiology has been one of the most labor- and skill-intensive sections of the clinical laboratory. An attempt to provide rapid microbiology information has been hindered by the need to obtain growth on conventional primary isolation media, (usually requiring at least 18 hr) prior to organism identification and antimicrobial susceptibility testing. In most cases, information has not been available until 36 to 48 hr after specimen receipt. Because of this prolonged delay, the data have been of limited use to the physician. Automation has in part attempted to shorten the time required for microbiological results. In some cases, the application of automation to the clinical specimen has not only shortened the detection time, but has eliminated the use of primary isolation media.

Numerous publications cited in this text and elsewhere report on the high degree of accuracy of the automated methods. Most of these automated systems correlate with reference methods at a level of ≥90%. Accuracy of automated results is less subject to misinterpretation than manual methods. Instrument-printed test results reduce errors associated with manual record keeping and data handling. Automation thus increases productivity and tends to reduce human error related to repetitive and monotonous work. Increased productivity should allow skilled microbiologists more time for performance of more demanding analyses. Routine procedures that previously required the full commitment of microbiologists can now in some cases be accomplished by an instrument.[3]

A major charge has recently been placed upon all hospital personnel to participate in lowering costs in order to survive under the new system of third-party prospective medical payments in the U.S. (the diagnosis-related group or DRG system). Under this system, the hospital is encouraged to shorten the length of the patient's stay. Pressure is being placed upon the laboratory to shorten the turnaround time of test results in order that the appropriate DRG may be established early in the hospital admission. It is a well-recognized fact that the major expense within the laboratory is labor. it has been stated that labor reductions are best accomplished through attrition and increased automation.[4]

DRGs have been in effect for over 3 years in all acute-care hospitals in New Jersey. One New Jersey microbiologist responded to the changing climate by performing automated microbial identifications and antimicrobial susceptibility testing in order to obtain results within 4 to 6 hr instead of overnight.[5] Additional reduction in turnaround time was accomplished by using a urine screening instrument that provided final negative urine reports within minutes for approximately 60% of the specimens. Automation has improved turnaround time and workload efficiency under DRGs in this New Jersey institution, thus allowing the lab to expand into other diagnostic areas such as herpes and chlamydia testing. MIC susceptibility testing and reporting by an automated method provided a more cost-effective use of antibiotics and, in turn, this provided a recognized benefit to the pharmacy and its budget.[5]

Automation has contributed to increased accuracy of bacterial identification to the species level. Prior to the 1960s, microbiologists relied on a few biochemical tests for the species identification of microorganisms. The selection of these tests was dependent upon the availability of the media within the laboratory as well as the individual performing the procedure. The use of a small number of tests limited organism identification and unusual species may not have been recognized. Performance of large numbers of tests was not practical or cost-effective. Interpretation of test results was accomplished through the use of taxonomic tables showing the percent positive for each test.[6] Final identification was subjective and certainly dependent on the skills and knowledge of the individual interpreting the test reactions and identification scheme. The introduction of commercial identification systems during the 1960s provided several advantages over the previous system. These multi-test commercial systems allowed for the identification of common as well as unusual species, provided extended shelf life of reagents and biochemicals, introduced some consistency within the laboratory, provided ease of inoculation, and introduced computer-assisted interpretation through the computer generation of profile code numbers or biotypes. Although some systems were rapid, interpretation of results was still dependent upon the individual reading the reactions. For rapid results (4 to 5 hr), it has been necessary for tests to be set up early in the work day to allow reading of results prior to the end of normal working hours.

Present-day automated identification systems have incorporated many of the concepts of the manual commercial systems with further advancements. Additionally, they provide objectivity, potential labor savings, the ability to complete tests beyond usual work hours, expanded identification capabilities, and improved epidemiological data. Many repetitive tasks have been eliminated through the use of automated identification systems. These include preincubation, pipetting, oil overlays, insertion of discs, addition of reagents, interpretation of test reactions, recording of results and use of taxonomic code books. Another advantage is that both the identification and susceptibility testing can often be performed from the same broth or saline sample. Elimination of these tedious tasks has contributed to both time and labor savings.

Possible guidelines for acceptable accuracy of automated bacterial identification and antimicrobic susceptibility tests have been suggested.[7] Numerous reports have shown that these criteria have been met. For bacterial identification, overall correct species level identifications approximate at least 90% and secondary species identifications are suggested when identifications of lower statistical confidence occur. Additionally, some of the systems provide a preliminary identification within 4 hr followed by complete definitive identification after a few additional hours of incubation of the same subculture without further cost.

The automated antimicrobial susceptibility test systems probably have been evaluated more than any other test procedures in the microbiology laboratory. Results indicate that the percentages of very major and major errors are minimal, pathogens with unusual patterns of susceptibility are usually detected, and warnings of possible problems are indicated by some test systems. Expanded testing of newer antimicrobics generally have become available in a timely manner. Perhaps, the most technically pleasing example of automation in microbiology is the assay of antimicrobial agents. Prior to automation, the bioassay method was used to determine antimicrobic concentrations in body fluids. Although this method has the advantage of employing a technique which is familiar to most microbiologists, there are many disadvantages. Bioassays are limited in the range of antimicrobics which can be tested, interfering substances including the presence of other antimicrobics can effect test results, the procedures are somewhat tedious and time-consuming to perform, they are less accurate and precise than more recent instrument methods, and they have a rather lengthy turnaround time (4 to 6 hr).[8] Automated assay methods, previously discussed in this text, have eliminated virtually all of the these difficulties. As a result of automation, assays of antimicrobial agents are now sensitive, accurate, simple to perform, and results may be available within minutes.

A major accomplishment of automation in microbiology has been the introduction of microcomputers with the automated systems. These microcomputers not only analyze individual test results, but also sort and assemble complex laboratory data, such as epidemiological data, which was previously performed manually. While the rapid generation of test results is a significant accomplishment, this information must reach a physician as rapidly as possible to have real meaning. This is possible by interfacing instruments to a laboratory or other host computer for instantaneous transfer of information. However, a problem occurs when test results become available during the hours that the microbiology staff is not present for result verification. This obstacle has been overcome to a considerable degree by introducing edit programs in the laboratory computer that allow transfer of information to patients' files only if certain criteria are met.[9] For example if an organism identification has a >90% probability, it will be directly reported and then later reviewed by a technologist. This type of reporting system allows the microbiologist to enter data directly as information is generated. This interaction of automated microbiology systems with other computers has been a definite accomplishment.

Direct entry into an automated reporting system not only eliminates the need to transcribe data, but also decreases the possibility of errors encountered with a multistep process involving both technical and clerical personnel. Most importantly, direct data entry eliminates any intermediate delays in reporting of results.

In summary, the extensive accomplishments of automation in microbiology have had a direct beneficial effect on both the patient and the laboratory. Automation has allowed us to overcome one of our greatest obstacles, excessive turnaround time. We are now able to report the presence of some infectious agents within minutes to a few hours after specimen collection. Expensive and time-consuming conventional culture procedures at times may be eliminated by screening tests. The use of inappropriate or ineffective antimicrobial therapy should decrease as commercial companies develop still more advanced technology for diagnosis of infectious diseases and as the willingness of microbiologists to accept "non-traditional" methods continues to increase through use of automation in the diagnostic microbiology laboratory.

REFERENCES

1. Matsen, J. M., How rapid should rapid methods be?, *Clin. Microbiol. Newslett.*, 4, 164, 1982.
2. Dupont, P. F., Automation in microbiology: a physician's viewpoint, *Am. J. Med. Technol.*, 49, 323, 1983.
3. Isenberg, H. D., Clinical significance of automation in microbiology, in *Rapid Methods and Automation in Microbiology*, Tilton R. C., Ed., American Society for Microbiology, Washington, D. C., 1981, 96.
4. Miller, J. M., The impact of DRGs on hospital laboratories, *Clin. Microbiol. Newlsett.*, 6, 57, 1984.
5. Evangelista, A. T., Two year's experience with DRGs: a clinical laboratory perspective, *Clin. Microbiol. Newslett.*, 6, 79, 1984.
6. Edwards, P. R. and Ewing, W. H., Identification of Enterobacteriaceae, 3rd ed., Burgess Publishing, Minneapolis, Minn., 1972.
7. Sherris, J. C. and Ryan, K. J., Evaluation of automated and rapid methods, in *Rapid Methods and Automation in Microbiology*, Tilton, R. C., Ed., American Society for Microbiology, Washington, D.C., 1, 1981.

8. Pezzlo, M. T., Assay of antimicrobial agents, *Am. J. Med. Technol.*, 49, 779, 1983.
9. Ryan, K. J., The computer in microbiology: future applications in test performance and reporting, *Ann. N.Y. Acad. Sci.*, 428, 243, 1984.

Chapter 16

THE SHORTCOMINGS OF CURRENT AUTOMATION IN CLINICAL MICROBIOLOGY

Joseph L. Staneck

TABLE OF CONTENTS

I. INTRODUCTION

The realization that instrumentation has come to clinical microbiology can be had by a stroll through the exhibit area of any major microbiology meeting. That the application of machines to microbiology has been successful is evidenced by the increasing presence of instrumented devices in the hospital laboratory. That this effort to automate will continue is likely given the growing level of acceptance of mechanization by microbiologists and the increasing attention by industry to provide such devices. That the clinical microbiology laboratory of the future will be largely automated, however, is a question worthy of some debate and one whose answer will be influenced by the availability of instruments capable of satisfying the needs of our changing laboratory practices.

The purpose of this chapter is to present somewhat of a devil's advocate perspective on the phenomenon of "automation" in clinical microbiology. The reader is asked to view the following text as constructive inquiry with due recognition of the positive aspects of instrumentation as illustrated by the author's colleagues in previous chapters. It is hoped that the exploration of rationales, experiences, and opinions of the author and other microbiologists whose wisdom the author has solicited regarding automation will stimulate further thought on this subject.

II. GENERAL CONSIDERATIONS

The most frequently asked questions regarding automation in microbiology include: "Why has it taken so long to occur?", "Why does so little of automation deal directly with specimen material?", "Has it saved either time or money?" and even "Does automation actually exist in clinical microbiology?" *Webster's New World Dictionary* defines automation as "any system or method... using self-operating equipment, electronic devices, etc. to replace human beings in doing routine and repetitive work."[1] It would follow then that "automation" is a relative term and it exists in varying degrees dependent on the amount of human replacement achieved. Therefore, automation in the quantitative sense does indeed exist in clinical microbiology, but the term is sometimes used loosely. "Instrumentation" is perhaps a better term since many current instruments are, in fact, low on the relative scale of automation and fail to "replace human beings". Inspection of available instruments reveals that most focus on one, two, and possibly but rarely, more steps in a complex series of processes that must occur from the time of specimen receipt to the issuance of a final report. There are but a few examples in microbiology comparable to the level of automation commonly encountered in areas of laboratory medicine such as clinical chemistry where hands-off processing proceeds from specimen loading to final report. The automation chapter in the recent text *Clinical Chemistry* notes, "Economics suggest that instruments perform more than one kind of test; that is, the investment in automation and the labor of sample loading is not duplicated."[2] Most instruments in microbiology that perform more than one test (i.e., identification and susceptibility testing) require a duplication of sample loading and plasticware and thus erode labor and cost-efficiency.

It is useful to examine the inability of the majority of instruments in microbiology to analyze direct specimen material, as well as the time lag that has existed in the application of instrumentation to microbiology relative to other areas of laboratory medicine by examining the activities of the clinical chemistry laboratory. Instrumentation revolutionized the practice of clinical chemistry well over 2 decades ago. The clinical chemist's dilemma is one of "which", while the microbiologist is still asking "whether" with respect to the implementation of instruments.

The differences between chemistry and microbiology begin with the fundamental mission of each when responding to a test request. The chemist responds to a very specific question concerning one or more defined analytes in the specimen. "How much calcium, glucose, protein, etc. is in this specimen?" The task, more often than not, is directed toward the measurement of that analyte. Most importantly, the analyte at the start of the analysis arrives in a rather homogeneous liquid (serum, body fluid, or urine). The analyte, providing appropriate care of the specimen is exercised, is stable. With rare exception, the analyte needs no amplification, and if so, such is provided by some type of concentration step that can be applied within a short time interval. If separation of the analyte is required, specific physical/chemical processes can be applied. Detection of the analyte is accomplished with the application of reagents and the fulfillment of a criterion highly specific for that substance thus allowing a defined endpoint. Lastly, once detected and measured, the task is complete and the report is issued simple numeric form.

In contrast, the microbiologist is more often than not asked to respond to a very open-ended question. "What," the physician asks, "can you tell me about this specimen that will assist me in the diagnosis and management of what I think is an infectious process in my patient?" The search may be as narrow as the recognition of a single entity, such as group A β-hemolytic streptococci in a throat specimen, or as broad as to include any one of a wide variety of bacterial, mycobacterial, fungal, viral, or protozoan infectious agents. These "analytes" are dynamic, living entities, not infrequently accompanied by other biologically dynamic entities of more or less importance to the infectious process, the presence or absence of which is dictated by choice and collection of specimen. No other laboratory's processing area hosts the variety of specimen types, the heterogeneity of consistencies, nor the dubious specimen quality as is found in the microbiology laboratory. Recognition of the "analyte" most often demands biological amplification within any one of a number of nutritional, temperature, and atmospheric environments. Once recognized, significance may be determined by the somewhat arbitrary criteria of numbers, predominance, prior assessment of the specimen quality, and last but not least the very nature of the beast. Characterization of the analyte may include its staining properties, its activity upon red blood cells, its enzymatic complement, its ability to utilize specific nutrients, its possession of certain antigens, its interaction with antimicrobial agents, or any combination of the above and possibly more. Lastly, at the end of the analysis, a descriptive, sometimes qualified report is sent forth.

Given the above, there is little wonder that the application of either automated or, for that matter, manual technology has been so limited with respect to direct specimen analyses despite the clear need and desire of the microbiologist to provide information without the lag time of conventional culture. It is also not surprising that the most successful attempts to automate or employ instrumentation that deals directly with the specimen have been in situations where the test parameters resemble those found in the chemistry laboratory. In other words, a specific question is asked relative to a particular type of homogeneous liquid specimen. Successful instrumentation/automation identical to that of our chemistry colleagues can be found with the TDX® instrument (Abbott Laboratories, North Chicago, Ill.) or the EMIT Autolab system (Syva, Palo Alto, Calif.) for the determination of antibiotic levels in serum. In fact, the measurement of antibiotic levels is not infrequently performed within the chemistry or toxicology section, because of common technology. There is a disturbing implication in such arrangements that the knowledge, skills and expertise of the microbiologist in the characterization and interpretation of the interrelationships between the host, the microbe and the antimicrobial agent may be preempted by the abilities of other laboratory scientists to technically perform a test. It appears incumbent on a microbiologist to master developing technologies and to discourage the jurisdictional assignment of tests based not on scientific expertise but on equipment considerations alone.

The Bactec® instrument (Johnston Laboratories, Towson, Md.) for the monitoring of blood cultures owes it success to the ability to ask a specific question: "Is there sufficient CO_2 present in the atmosphere of this bottle to suggest microbial growth within the broth?" Once a question can be so specifically defined, the difficulty of the task of finding suitable technology to answer the question is reduced considerably. The detection of radiolabeled CO_2 has functioned well for this purpose, as will no doubt the application of the new nonisotopic, spectrographic CO_2 detection mechanism. Likewise, the area of bacteriuria screening lends itself to an instrumented approach because again, a rather specific question can be asked: "Are there a sufficient number of organisms present in this specimen to suggest urinary tract infection?" Unfortunately, as will be mentioned later, the question itself may be subject to redefinition. The point, however, remains the same. The more specific the question, the more likely a technology can be developed to provide an answer, and the greater the possibility that such technology can be mechanized or automated.

However, the microbiologist asks few specific questions. Even the newer technologies of nucleic acid probes and monoclonal antibodies are useful only when the search is narrow enough to be economically and logistically feasible. In addition, these technologies have yet to address the question of susceptibility testing. Therefore, the present shortcoming of current instrumentation to effectively deal with direct specimen material may be a future shortcoming as well. The culture plate may remain the most frequently used means of direct specimen analysis in microbiology.

Given the difficulties of dealing directly with specimens, it was reasonable that automation efforts be turned toward those tasks in microbiology that could be more specifically defined, most occurring after conventional culture had yielded an isolated colony of the suspect organism. Previous chapters detail these instruments, generally used to perform susceptibility testing and/or identification of bacteria.

To find where this array of hardware has fallen short of our expectations, the author solicited the opinions of a number of colleagues with experience using one or more of these systems. The author is indebted to each of the respondents for the giving of their time, effort and professional insight. A total of 21 laboratories (including the author's) was represented in this survey with hospital bed size ranging from 320 to 1200 beds, averaging 620. Approximately one half of this group was either associated or affiliated with a university, the remaining half being private community hospitals. All laboratories are now using or had in the past used some type of instrumentation. There were users, past or present, of: Bactec® (67%), General Diagnostic's Autobac® (48%), Vitek's AutoMicrobic® system or AMS® (43%), Abbott's MS-2® (21%), Gibco's Sensititre® (14%), American Micro Scan's Autoscan® (10%), BBL Sceptor® (10%) and Marion Laboratory's Bac-T-Screen® (5%). Curiously, no laboratory other than the author's listed instrumentation for aminoglycoside monitoring assays. (We have had experience both with the EMIT Autolab and the Abbott TDX®.) Whether this omission was through an oversight on the part of the respondents, the performance of these assays in laboratory sections other than microbiology, or the opinion that these instruments are not really microbiology instruments has not yet been determined. Application of the various instruments mentioned in the survey included susceptibilty testing (81%), blood culture (67%), identification (61%), urine screening (14%), mycobacteria detection (14%), data management, or word processing (14%).

When asked to rank various considerations of instrument purchase, accuracy and rapid turnaround time received the highest priority among the respondents. Sharing second place were the running costs or consumable costs per test of the instrument and the degree of labor savings offered, two related considerations as both affect lab budgets. Four parameters occupied third place priority: fixed or purchase price of the instrument; flexibilty of the system (customizing potential); ability to carry out more than a single type of test or function; and minimal disturbance of laboratory work flow. Holding fourth and last place in priority was the physical size of the instrument. Apparently, users did not care how big the instrument as long as it did the job quickly and accurately. Given these listings, it was therefore interesting that the same respondents when asked to cite the advantages of their respective purchased instruments mentioned increased accuracy only once (and this was in reference to a previously used instrument, not relative to a manual method). The majority of responses, however, did cite better turnaround time. Therefore, while both parameters appear to be equally important in the minds of the user before purchase, clearly in practice the user is willing to allow some trade off of accuracy (which appears to be less for automated vs. manual methods) for shorter turnaround times. The advent of prospective payment plans and the increasing pressure to provide more responsive laboratory services may, in fact, further this philosophical approach.

Cost-effectiveness was noted in the survey as a high-priority, desirable attribute. The respondents' concern seemed to be more with day-to-day operating costs than with the capitalization of the instrument itself, although these two parameters may converge in reagent-rental options where a payback of the capital cost is made through a taxation of reagent or plasticware purchase. It could be argued that cost-effectiveness is one area in which instrumentation in clinical microbiology has offered questionable progress. Few respondents reported that the cost of materials per test of an instrument approach was found to be less than the cost of the previously used manual method; in fact, the majority reported increases in cost with instrumentation. On the other hand, in consideration of the overall cost impact of employing instrumentation, 90% of Bactec® users, for example, felt that their system was indeed cost-effective due to savings in media and supplies formerly used to perform blind subcultures of blood bottles and in some cases to a labor reduction. Users of the AMS,® Sensititre® and Autoscan® also felt the systems to be cost-effective when taking into account both direct and indirect costs associated with running the tests. Autobac® and MS-2® users were less enthusiastic about cost justification, with only half of the respondents employing these instruments noting some degree of cost-effectiveness. (It is appropriate to remember in cost discussions that in some instances an increase in direct test cost may, through some attractive attribute such as decreased turnaround time, actually reduce total hospital costs.) In our own laboratory, the Abbott TDX,® with excellent standard curve stability, instrument reliability and virtually total automation has not only been cost-effective compared to a bioassay or manual antibody-linked "kits" but has permitted us to maintain a very active, responsive drug monitoring service with reduced staffing because of the high degree of productivity realized with this system.

Reverting back to the theme of "replacing humans", one theoretical goal of automation should be that of labor savings. How did current instrumentation measure up in this category? Not as well as one would have liked. Only half of the Bactec® users felt that some degree of reduced staffing occurred with adoption of the instrument. While the mechanical analog of blind subculturing of bottles was highly automated, the manual adjunctive tasks of logging, loading and checking out false-positives, etc. were time consuming. In some laboratories, the Bactec® allowed for more than the traditional subculture once a day on alternate days. While this ability to do more frequent monitoring was a distinct advantage with respect to detecting positives more rapidly, the extra labor involved cancelled that saved by machine "subculture".

AMS®, a highly automated instrument for identification and susceptibility testing (once the plasticware is loaded), in the majority of the laboratories responding either allowed the substitution of technician for technologist time or did directly effect labor reduction. Estimates ranged from 8 to 20 hr per week among AMS® users. Few users of the Autobac® or MS-2® realized labor savings and several respondents commented on cumbersome preparation procedures. Users of Sensititre® and Autoscan® tended to report an increase in time savings as more mechanized equipment was applied to the microdilution format compared to the time spent on carrying out microdilution assays without instrument assistance.

III. SPECIFIC CONSIDERATIONS

It would perhaps be convenient at this point to turn the discussion from generalities to specific tasks performed by various instruments and to examine the processes with a judgmental eye. Automation of the detection of microorganisms in blood is available only with the Bactec® system. In terms of acceptance by laboratory technologists, the Bactec® averaged 8 on a scale of 1 to 10 within our survey (10 being the greatest acceptance). This was the highest level of acceptance by the laboratory staff of any instrument reported. However, several drawbacks surfaced among the survey responses. Concern was expressed that the relatively small volume of blood accommodated by the Bactec® bottles might possibly lead to a loss of detection in situations where organism density was very low. Evidence does suggest that the more blood drawn for culture, the more likely the recovery of organisms, particularly Gram-negative bacilli.[3] Also, the aforementioned tradeoff in labor due to more frequent sampling to optimize recovery was felt to negate major time savings. The limited capacity of the Bactec® 460 also necessitates some degree of bottle shuffling particularly in large-volume laboratories. It should perhaps be noted here that a new generation Bactec®, using nonisotopic technology, is more highly automated than the current Bactec® 460 series and possesses a greater capacity for bottle storage enabling the instrument to perform many days of sampling in a totally hands-off fashion.

A more fundamental concern might be to question whether recent improvements in manual blood culturing systems have diminished the advantages of automation. Both the Isolator® (DuPont, Wilmington, Del.), and the commercial biphasic blood culture systems eliminate routine blind subculture of bottles and provide not only the detection of organisms comparable in time to Bactec®, but in many instances yield next-day isolated colonies upon which rapid identification and susceptibility procedures may be carried out. In light of these systems, the major shortcoming of the Bactec®, therefore, may be that it is capable only of performing the first step in a multi-step process, detection. Although the Bactec® does this single step well without human intervention, the visual-only inspection of the manual aerobic biphasic blood culture bottle minimizes the labor of the detection step and allows the second step of isolation by subculture to be accomplished by a simple tilt of the bottle. Pfaller et al.[4] have reported excellent recovery of organisms from blood plus next-day colony recovery in over 60% of positive specimens using the Septi-chek® (Roche Diagnostics, Nutley, N.J.). A recent study in the author's laboratory has generated comparable data using the biphasic bottle of Gibco Laboratories.[22] In both studies, the simple manual processing allowed for an advantageous evening subculture the day of receipt. In addition, data from my study plus the reports of others[5,6] question the value of blind subculture of macroscopically negative anaerobic monophasic bottles. Therefore, the use of an aerobic biphasic-anaerobic monophasic blood bottle set may require only visual inspection without blind subculture. Although the cost per bottle may be less for the Bactec® than for either the Isolator or the biphasic bottle, avoidance of capital expenditures and equipment maintenance fees, plus early colony growth make the nonautomated methodologies attractive.

Among the instrumented approaches to antimicrobial susceptibility testing, the Autobac® has the longest history. This innovative system heralded the era of instrumentation in microbiology a decade ago and made the concept of same-day susceptibility testing a practical reality. However, the survey information indicated that the Autobac® was the most discontinued instrument among all reported. In fact, among the respondents, seven of ten Autobac® users have adopted alternative methodology to carry out susceptibility testing. Curiously, four laboratories shifted to same-day automation with AMS®, two laboratories shifted to next-day instrument-assisted microtiter systems, and one laboratory chose to return to a manual next-day system. Why these departures from Autobac®? Four respondents sought a greater degree of automation than the Autobac® had to offer. Three laboratories perceived accuracy problems that apparently prompted a return to more conventional susceptibility test modalities. Several reports in the literature suggest specific organism-drug combination problems with the Autobac® such a methicillin-resistant *Staphylococcus aureus*, aminoglycoside-resistant *Pseudomonas aeruginosa*, and others.[7-10] Recently, in the author's laboratory, 30% of *P. aeruginosa* giving gentamicin-sensitive Autobac® I results were found to have MICs of 8 $\mu g/m\ell$ or greater by microbroth dilution testing in cation-supplemented Mueller-Hinton media, thus placing these isolates in the intermediate or resistant categories. Other users, including my own laboratory, have through the years discontinued the testing of certain organism groups on the Autobac® because of accuracy concerns. The next most frequent criticism involved the long delay in the approval of newer antibiotics for the Autobac®, thus limiting its application in an era of novel and changing antibiotic usage. Lastly, several respondents felt that the labor component of the Autobac® was in the range of manual methods and therefore offered no advantage. The MIC testing on the Autobac® has logistic limitations and requires either duplicate or expanded plasticware. Perhaps for the above reasons the acceptance of the Autobac® instrument by the laboratory staff was rated at 5 (average) on a scale of 10.

Although the MS-2® (now called the Avantage) was less represented in the survey, similar comments concerning the time involvement of specimen preparation and the delay of new antibiotic capabilities were also noted. There appear to be problems in the accuracy of results for certain drug-organism combinations with the MS-2®, not unlike other same day systems.[10-13] Perhaps the greatest disappointment to the MS-2® user was that the time savings initially hoped for failed to be actualized.

The AMS® (Vitek) is a system which has automated conventional susceptibility testing and identification processes. Although AMS® users at present seem to be more satisfied with their instruments than in the past, perhaps attributable to the continuing development and technical service of the manufacturer, certain problems with susceptibility testing have been reported.[10,13,14] The "same day" AMS® susceptibility results, in practice, may require up to 8 or 10 hr, and probably average 5 to 6 hr. This interval imposes work flow restrictions and diminishes the impact of "same day" testing if in fact data cannot be processed and transmitted until the next day. Acceptance by technical personnel nevertheless was rated 7 on a scale of 10. Far and away the greatest drawback mentioned by the AMS® user is the necessary use of fixed configurations in the plasticware. Novel drugs or custom panels are not readily available and time will determine how important these factors will become to widespread acceptance of the system. The initial expense of the AMS® is also a drawback and may limit the use of this instrument to the larger laboratory.

The microtiter tray format for susceptibility testing has the advantage of flexibility, uses conventional methodologies and lends itself well to instrument accessory. A variety of commercially prepared microtiter susceptibility trays with adjunctive inoculation and/or reading devices can be found in the marketplace and have been discussed in earlier chapters. Every major contender in the microtiter market offers some type of computer assistance in reading, data recording and interpretation, report printing, and subsequent data management. These devices qualify only minimally as automation in that while they have distinct advantages for the user, they require operator presence and do not really "replace humans". However, progress should be noted in some areas. The automatic inoculation and hydration devices for the dehydrated systems (Gibco Sensititre®, API UniScept®, BBL Sceptor®) are almost a necessity for efficient routine use of these products, saving both the labor and frustration of pipette inoculation. All three systems offer instruments that actually read MIC endpoints or reactions in microtiter plates. The Sensititre® Autoreader is capable of reading standard MICs through the use of innovative technology employing fluorogenic enzyme substrates. The Autoscan® and UniScept® instruments are capable of reading both MICs and biochemicals for the identification of Gram-negative bacilli. None of these automated readers at present operate in a "hands-off" mode. The instruments currently function after a single plate manual feed. Clearly, if labor savings is an objective, the mechanical configuration of these instruments must allow for the stacking of microtiter trays and subsequent reading without operator assistance. Plans to this effect are being made by several manufacturers.

The most frequently mentioned drawback of the existing microtiter systems is that results require overnight incubation. This standard incubation period using conventional antibiotic dilution patterns (or biochemical reactions) does allow for excellent accuracy compared to macro-broth conventional procedures, but is not considered in the "rapid" test category. After an 18-hr incubation, automatic tray readings of both the Sensititre® Autoreader and the Autoscan® are reported to reflect manual readings of their respective microtiter plates,[15,16] The novel fluorogenic enzyme substrate technology of the Sensititre® Autoreader has allowed for the development of an MIC system that can be instrument-read and interpreted at 4 to 5 hr of incubation. Collaborative ongoing studies in the author's laboratory and those of two colleagues (Dr. S. D. Allen, Indiana University, and Dr. R. C. Tilton, University of Connecticut) suggest that this will be a feasible same-day microtiter system for susceptibility testing of Enterobacteriaceae.[23] This same technology is currently under development for bacterial identification, and it would therefore appear that at least one major system is dedicated to overcoming the next-day data drawback of present microtiter applications.

Instrument approaches to identification of facultatively anaerobic and glucose nonfermenting Gram-negative bacilli are possible using AMS, MS-2® and the Autobac®. A major shortcoming of these systems mentioned earlier with respect to susceptibility testing included the inflexibility of drug configurations, not a concern with identification systems. Therefore, the pertinent issues become accuracy and labor savings. While accuracy appears to have improved with development, labor savings over manual methods is variable and often minimal. Again, to ask a more fundamental philosophical question, does one need automation for identification purposes given the reasonable cost and ease of use of manually read multi-reagent kits? With respect to Enterobacteriaceae and *Pseudomonas,* many laboratories have adapted the useful and very inexpensive shortcuts of spot indole for *Escherichia coli* and swarming *Proteus* identifications, and cytochrome oxidase, pyocyanin production for *Pseudomonas aeruginosa,* thus obviating the need to use the more expensive automated or even manual multi-reagent methods for a fair portion of Gram-negative isolates. All major microtiter systems also offer accurate identification systems for gram negative organisms.

Some offer instrument reading while most offer options for computer-assisted reading, data base interpretation, and report printing. These systems require overnight incubation, again generally considered by be a drawback. The material and labor cost of these instrumented and instrument-assisted identifications may be comparable to and in some cases exceed that of manual "kits". Advantages of use, therefore, tend to be continuity of work flow if the instrumented systems are utilized for other laboratory testing, and the utilization of the microcomputer components for profiling, data management, report printing, and epidemiological analyses.

Lastly, let us look critically at the instrumentation applied to urine screening. The industry has directed its efforts toward the direct determination of bacteria in urine that have a density of approximately 100,000 colony-forming units (CFU) per milliliter or more, a time-honored discrimination point between "significant bacteriuria" and "contamination." However, the work of Stamm and his collagues[17] has provided compelling evidence that numbers of Gram-negative bacilli less than 10^5 CFU/m*l* may be quite relevant clinically, particularly in symptomatic females. It is not so important here to resolve the debate on this issue as it is to recognize that this debate exists. Confusion over interpretation based on numbers of organisms in urine is likely to increase in the future. Requirements for increased sensitivity in certain patient groups may exceed the detection abilities of current systems. Such a state of affairs bears directly on the usefulness of urine screening at 10^5 CFU/m*l* as a practical tool. Let us assume for purposes of this discussion, however, that application of the 10^5 number as a discrimination point is useful, if only in particular settings, and ask how existing instrumentation performs. In this context, one can think of two major motivations for the screening of urine: to provide a rapid presumptive diagnosis (i.e., bacteriuria or not?); and to determine the need to perform conventional culture on the specimen. Several approaches utilizing instruments exist including turbidometric changes in broth diluted urine following incubation periods of 3 to 6 hr or more (Autobac®, MS-2®, AMS), the determination of levels of bacterial-associated ATP using firefly luciferinase technology (3M's Lumac; Analytical Luminsecence Lab's Microscreen) and the filtration-colorimetric determination of organisms (Marion's Bac-T-Screen®). Details of these systems can be found in an earlier chapter of this book. Assuming an acceptable performance of each in the discrimination of "bacteriuria" from "contamination" at 10^5 CFU/m*l*, what shortcomings exist?

Whether the intent is to provide a diagnosis or to decide whether or not to culture, those approaches which are based on an incubation time of three or more hours immediately impose work flow limitations, particularly in the single-shift laboratory because the specimen must enter the screening process early to be interpretable within the working day. In our own laboratory, where urine specimens are received around the clock, and many outpatients clinics meet in the afternoons, an incubation dependent screen is of limited value even with second-shift staffing. Systems such as the Autobac® and MS-2®, while low in cost, also require the refrigerated storage and retrieval of urine specimens (for possible culture), a cumbersome process for a large volume laboratory. The AMS® urine screen is unique in that beyond the screening process, the instrument can often proceed to an identification automatically giving a next day final identification report. However, the system does this at substantial expense, if one assumes that a high percentage of specimens may be "negative". Subculture of the card is also required for susceptibility testing. Some users apply a prescreen technique to urine before an AMS® urine card is set up in the interest of cost containment. One of the more popular prescreen methods is the determination of leukocyte esterase using a Chemstrip® LN (Biodynamics, Indianapolis, Ind.). The fundamental question thus arises whether this type of very rapid, inexpensive analysis for pyuria might replace the microbiologically-oriented methods when screening certain patient populations.

The innovative ATP detecting technology using firefly luminescence to monitor the number of microbial cells in urine has the advantage of rapid turnaround time, but is labor intensive and uses regents of limited shelf life and therefore, is costly. Manufacture of the Lumac instrument has been discontinued, presumably due lack of interest in the market place. The Microscreen kit (Analytical Luminescence Lab) using a Monolight 500 instrument addresses some of the earlier drawbacks of this technology by simplifying the procedure. The success of the ATP detection concept will depend on the ability of industry to offer practical, low-cost kits that are simple to use.

The 2-min turnaround time and simplicity of the Bac-T-Screen® make it a seemingly attractive option for urine screening. Several studies[18-20] have been favorable in terms of the system's ability to categorize specimens as significant bacteriuria or not at the 10^5 CFU/mℓ level. The biggest drawback appears to be the cost per specimen, which tends to be close to the cost of conventional quantitative loop culture, no matter how cleverly the cost accounting is carried out. If the system is used only to decide which urines to culture, the user pays as much for the screen than for the procedure it is designed to eliminate. Therefore, a premium is being paid (and depending on the circumstances and viewpoint of the user, possibly a worthwhile premium) for the immediate answer. This instrument, which is simple to operate, analyzes a single specimen at a time and in practice must be operator attended. However, a second generation instrument offers a reduction of the present 2-min dwell time to 1-min and further automation seems unwarranted in light of such a short turnaround time. The reports mentioned above note that pigmented specimens and specimens which clog the filter pad of the detection card may invalidate approximately 12% of specimens, a problem that presents both logistic and economic concern. On the positive side, a recent report suggests that the Bac-T-Screen® has a reasonable sensitivity for detecting either high numbers of bacteria alone or lower numbers (down to 10^3 CFU/mℓ) of bacteria in the presence of pyuria, therefore, may be useful to identify specimens needing no further microbiological workup.[21]

In general, the drawbacks of present systems that screen for bacteriuria include (1) discrimination ability based on counts of 10^5 CFU/mℓ, a concept subject to controversy, (2) performance characteristics that favor negative predictive values or "ruling out" ability, (3) material, instrument and labor costs that rival that of conventional culture which must be performed in addition on screen-positive urines, and (4) long, work flow-limiting incubation periods with growth-dependent systems.

IV. SUMMARY

The major shortcoming of instrumentation in clinical microbiology appears to be the limited ability of existing systems to deal directly with specimen material. While the advantage of same-day turnaround time would be considerable, the dynamic, multifaceted, often ill-defined nature of the typical microbiological analysis presents formidable obstacles to a cost affordable resolution of this deficit. Those approaches that deal with direct analyses of specimens are either one dimensional in function (automated blood culture) or if multi-functional (automated urine screen with identification), tend to be costly. The major exception to this is perhaps the instruments that determine antibiotic levels which, like their counterparts in chemistry, represent a high degree of automation and allow for a significant savings in labor.

Among the popular instruments that perform susceptibility testing and identification once an organism is isolated, concern exists over the lack of same-day testing with some, the accuracy of the rapid test performed by others, and the inflexibility of certain systems to allow for user variation, particularly in the selection of antibiotic panels. Generally, there is disappointment that labor savings have been minimal compared to previously used manual procedures and have not begun to approach the reduction of labor associated with instruments used by other areas of laboratory medicine. There also seems to be concern that many instrumented systems dress up conventional processes with attendant cost increases and questionable advantages.

In conclusion, a certain degree of success and acceptance has been realized with current instrumented approaches. However, amid the fanfare, whistles, and bells of present instrumentation, benefits greater than those currently achieved are being sought, particularly in the areas of labor savings, more rapid turnaround time and user customization. The challenge to industry is to continue to improve upon these parameters and to offer useful and affordable products. The challenge to the microbiologist is to choose the tools of his discipline, instrumented or not, with wisdom and to apply these tools in ways relevant to patient care and sound laboratory management.

REFERENCES

1. *Webster's New World Dictionary,* 2nd college edition, World Publishing, New York, 1970.
2. Davis, J. E., *Automation in Clinical Chemistry, Theory, Analysis and Correlation,* Kaplan, L. A. and Pesce, A. J., Eds., C. V. Mosby, St. Louis, 1984, chap. 15.
3. Hall, M. M., Ilstrup, D. M., and Washington, J. A., Effect of volume of blood cultured on detection of bacteremia, *J. Clin. Microbiol.,* 3, 643, 1976.
4. Pfaller, M. A., Sibley, T. K., Westfall, L. M., Hoppe-Bauer, J. E., Keating, M. A., and Murray, P. R., Clinical laboratory comparison of a slide blood culture system with a conventional broth system, *J. Clin. Microbiol.,* 16, 525, 1982.
5. Murray, P. R. and Sondag, J., Evaluation of routine subcultures of macroscopically negative blood cultures for detection of anaerobes, *J. Clin. Microbiol.,* 8, 427, 1978.
6. Paisley, J. W., Rosenblatt, J. E., Hall, M., and Washington, J. A., Evaluation of a routine anaerobic subculture of blood cultures for detection of anaerobic bacteremia, *J. Clin. Microbiol.,* 8, 764, 1978.
7. Stubbs, K. G. and Wicher, K., Laboratory evaluation of an automated antimicrobial susceptibility system, *Am. J. Clin. Pathol.,* 68, 769, 1977.
8. Cleary, T. G. and Mauer, D., Methicillin resistant *Staphylococcus aureus* susceptibility testing by an automated system, Autobac I, *Antimicrob. Agents Chemother.,* 13, 837, 1978.
9. Mayo, J. B., Kiehn, T. E., Wong, B., Bernard, E. M., and Armstrong, D., Examination of *Pseudomonas aeruginosa*-gentamicin discrepancies encountered in an Autobac I-disk diffusion comparison, *Antimicrob. Agents Chemother.,* 21, 412, 1982.
10. Hansen, S. L. and Freedy, P. K., Concurrent compatibility of automated systems and commercially prepared microdilution trays for susceptibility testing, *J. Clin. Microbiol.,* 17, 878, 1983.
11. Boyce, J. M., White, R. L., Bonner, M. C., and Lockwood, W. R., Reliability of the MS-2 system in detecting methicillin resistant *Staphylococcus aureus, J. Clin. Microbiol.,* 15, 220, 1982.
12. Carlson, J. R., Conley, F. E., and Cahall, D. L., Methicillin-resistant *Staphylococcus aureus* susceptibility testing with the Abbott MS-2 system, *Antimicrob. Agents Chemother.,* 21, 676, 1982.
13. Johnson, J. E., Jorgensen, J. H., Crawford, S. A., Redding, J. S., and Pruneda, R. C., Comparison of two automated instrument systems for rapid susceptibility testing of gram negative bacilli, *J. Clin. Microbiol.,* 18, 1301, 1983.
14. Woolfrey, B. F., Lally, R. T., Ederer, M. N., and Quall, C. O., Evaluation of the AutoMicrobic system for susceptibility testing of *Pseudomonas aeruginosa* to gentamicin, tobramycin and amikacin, *J. Clin. Microbiol.,* 19, 502, 1984.
15. Staneck, J. L., Allen, S. D., Harris, E. E., and Tilton, R. C., Automated reading of MIC microdilution trays containing fluorogenic enzyme substrates with the Sensititre Autoreader, *J. Clin. Microbiol.,* 22, 187, 1985.
16. DeGirolami, P. C., Eichelberger, K. A., Salfity, L. C., and Rizzo, M. F., Evaluation of the AutoScan-3 a device for reading microdilution trays, *J. Clin. Microbiol.,* 18, 1292, 1983.
17. Stamm, W. E., Counts, G. W., Running, K. R., Fihn, S., Turck, M., and Holmes, K. K., Diagnosis of coliform infection in acutely dysuric women, *N. Engl. J. Med.,* 307, 463, 1982.
18. Pezzlo, M. T., Wetkowski, M. A., Peterson, E. M., and de la Maza, L. M., Evaluation of a two-minute test for urine screening, *J. Clin. Microbiol.,* 18, 697, 1983.

19. Davis, J. R., Stager, C. E., and Aroj, G. F., Clinical laboratory evaluation of a bacteriuria detection device for urine screening, *Am. J. Clin. Pathol.*, 81, 48, 1983.
20. Pfaller, M. A., Baum, C. A., Niles, A. C., and Murray, P. R., Clinical laboratory evaluation of a urine screening device, *J. Clin. Microbiol.*, 18, 674, 1983.
21. Daifuku, R., Zeleznik, D., and Stamm, W. E., Detection of bacteriuria and pyuria with the Bac-T-Screen bacteriuria detection device, in Abstracts of the 1984 ICAAC, p. 280, #1072.
22. Weckboch, L. S. and Staneck, J. L., Performance characteristics of a commercially prepared biphasic blood culture bottle, *J. Clin. Microbiol.*, 23, 700, 1986.
23. Staneck, J. L., Allen, S. D., and Tilton, R. C., Rapid MIC determinations using the Sensititre Autoreader, in *Abstracts of the 1985 American Society for Microbiology*, C89, 314, 1985.

Chapter 17

PHYSICIAN ACCEPTANCE AND APPLICATION OF RAPID MICROBIOLOGY INSTRUMENT TEST RESULTS

James H. Jorgensen and John M. Matsen

Automated microbiology test devices do not necessarily provide results more rapidly than use of manual test systems; however, several of the automated instruments described in this text allow rapid reporting of important diagnostic information. The traditional, time-consuming practice of in vitro cultivation of microorganisms for diagnosis of infectious diseases has often been criticized for being excessively slow in providing information. Results generated by growth-based microbiologic methods to determine organism identification and antimicrobial susceptibility are often merely confirmatory of initial clinical impressions and drug selection. Susceptibility test results may be used only to reassess ongoing therapy rather than guide its initial selection.[1] This may be a contributing factor to improper or excessive antimicrobial use. One study based upon a retrospective chart review of physician prescribing habits noted that antibiotic therapy was inappropriate >40% of the time in hospitalized patients.[2] These authors noted that therapy was no more likely to be appropriate in life-threatening infections than in those that were less serious. However, they noted that the results from one rapidly available microbiology test (the Gram stain of sputum) was highly useful in guiding appropriate therapy of presumed respiratory infections.[2] It seems reasonable that microbiology results would be of greater impact if they were available at the time that initial therapeutic decisions were made.[3]

An obvious setting for use of rapid results is a medically urgent situation which requires prompt and accurate diagnosis to prevent serious morbidity or mortality.[4] Acute bacterial meningitis and other central nervous system infections represent medical emergencies in which prompt, accurate diagnosis is essential to prevent death or permanent neurologic disability. Infective endocarditis or bacteremia are potentially fatal conditions which also require prompt intervention. A number of studies have demonstrated that mortality from bacteremia can be sharply reduced through use of appropriate antibiotic therapy.[5-7] Conversely, inadequate to inappropriate antimicrobial therapy has been shown to triple the risk of death from bacteremia.[7] Other medically urgent situations include pneumonia in immunocompromised hosts, severe soft tissue infections such as necrotizing fasciitis, or burn wound sepsis, as well as numerous other invasions of deep tissues and normally sterile body cavities. While it has been difficult to document that currently available rapid susceptibility test results improve overall patient outcome,[8] the earlier provision of results in medically urgent situations may significantly improve selection of appropriate antibiotics.

The advent of prospective payment based upon DRGs for inpatient medical care in the U.S. places greater emphasis on efficient patient management in order to reduce the duration of hospitalization. It would seem reasonable that the more rapid availability of microbiology results through the use of automated instruments could contribute to a reduction in the length of hospital stay for treatment of infectious diseases. One of us (Matsen) documented that 9% of patients were discharged 1 day earlier when their physicians were aware of their same-day antimicrobial susceptibility test results.[1,3] These data were derived from a questionnnaire study conducted in 1976 based on 300 inpatients with infections. The physicians responded that with 95 patients (31.7%), they started a different antimicrobial agent upon receipt of the rapid preliminary culture and susceptibility results than they would have otherwise. With an additional group of 64 patients (21.3% of the total group) with serious or systemic infections, physicians indicated that therapy was changed to different antimicrobials based upon the rapid susceptibility results.[3] Furthermore, the physicians indicated that the length of hospital stay was shortened for 41 patients (13.7%). When this figure was corrected for underlying conditions which would by necessity prolong hospitalization, it appeared that 27 patients (9.0%) had their stay shortened because of the specific provision of earlier susceptibility results.[3] Such a questionnaire study has its limitations, but it provides encouraging evidence that rapid procedures may facilitate better, more efficient patient care.

It is also probable that results produced using rapid automated test instruments could contribute to improved diagnosis and management of outpatients with infectious diseases. If definitive laboratory tests could be completed very promptly, a patient's assessment could be completed during the initial examination period in the physician's office or clinic. This could reduce administration of unnecessary antimicrobials, and eliminate the need for patients to return for a second visit following the availability of the laboratory results.

Several common outpatient infectious diseases are amenable to this approach, including group A streptococcal pharyngitis, urinary tract infection, infectious diarrhea, and certain sexually transmitted diseases. A few rapid diagnostic tests are currently used for some of these processes, e.g., instruments for bacteriuria screening and gonorrhea and chlamydia detection.

The interest of clinical microbiologists in rapid test results has been largely restricted to the methods and instruments themselves, and thus restricted to only intralaboratory events. However, the total turnaround time for a laboratory test begins when the test is ordered by a physician and ends only when the test results are in the hands of a physician who can take action directed by the results of the test. The other phases of microbiology tests necessarily include specimen acquisition and result reporting in addition to the intralaboratory analysis. Far too little emphasis has heretofore been placed on efficient specimen acquisition and rapid result reporting. An earlier study noted that microbiology results were often not utilized by physicians in their selection of antimicrobial therapy.[2] In that setting, antimicrobial therapy was deemed appropriate 33.6% of the time before susceptibility data were known, and only 47.1% of instances after identification and susceptibility results were available.

As microbiology laboratories establish the capability of producing rapid results, the physician users must be informed about the availability and capabilities of the new tests. This can be accomplished using a variety of approaches such as laboratory newsletters, memoranda, or didactic presentations to the clinical staff. The laboratory's test request forms should be viewed as "menus," and thus should clearly indicate the availability of rapid instrument tests. The request form should not only obtain needed information relating to the patient and to the specimen, but should also provide the physician with an expectation of the time required for test results on that specimen. Far too often, physicians do not know when to anticipate the results of microbiology tests.

A lack of service at night or on weekends may erode the conditioning of physicians toward use of rapid results, and thus should be specifically considered in the overall staffing strategy of the laboratory. Thus, certain key tests for rapid diagnosis should be made available whenever possible to be performed in the "STAT" laboratory during the microbiology laboratory's off hours.

It is likely that the test results which most influence patient management are those which are available the earliest.[3] Thus, while the intralaboratory time required for a test should be as rapid as possible, additional efforts must be made to improve the method and timing of result reporting to physicians. Efficient use of results requires that their availability coincide with a physician-patient encounter, i.e., be available prior to physicians' inpatient rounds, or prior to an outpatient departing from a clinic or physician's office.

In a previously reported study,[1] the impact of 147 antimicrobial susceptibility reports on patients who had had no previous positive cultures was analyzed by contacting the respective physicians 72 hours after the reports were placed in the patients' charts. At that time, only 51.7% of the physicians were aware of the susceptibility results on their patients. In the 76 instances in which the physicians were knowledgable of the results, therapy was inconsistent with the susceptibility results in only 2 patients (2.6%).[1]

Laboratories must take a more aggressive posture in reporting results of microbiology tests. While direct telephone conveyance of urgent test findings to physicians is employed in many institutions, other rapid means of information processing must be pursued, i.e., computerized reporting systems. Certain of these systems are described elsewhere in this volume. It seems probable that until peripherals of a hospital-wide computer system are readily available in patient care areas, communication of laboratory results will suffer unnecessary delays. Physicians must be conditioned to expect the results of rapid tests, i.e., they must be aware of the likely turnaround times of particular tests and when the results should be forthcoming. If such results are provided only intermittently, physicians may be dissuaded from using the results of rapid tests to guide therapy.

Specific guidelines for the rapidity of urgent test results should use as a goal the provision of results within 1 hr of the time that the specimen was obtained from the patient. If results are not available within 1 hr, it is likely that physicians will use that information only to modify subsequent therapy rather than to guide or direct its initiation. Moreover, all less-urgent rapid microbiology results must be conveyed within the time period of physicians' normal working hours. For most likely utilization by physicians, results should be available in patient care areas by approximately 4 p.m. and not later than 5:30 p.m.

The laboratory must also engender in its physician users the confidence that the results of rapid test methods are suitably accurate. When a laboratory institutes new procedures aimed at providing rapid results, confidence must be instilled in the new methodology. In many cases, physicians will not realize that a laboratory procedure has been changed. Therefore, the accuracy and reproducibility of the new method must be described, including how the results of the newer rapid methods correlate with the results of traditional tests. This is particularly important in tests used in critical care patient situations. Trust in the new methodology can be engendered by a careful educational effort by the laboratory or by a clear correlation of the results with those from tests which physicians have been accustomed to receiving. Thus, there is need for careful developmental work within each laboratory to assure that any new test method functions as it should in individual settings, so that physician confidence in test results will not be eroded.

Certain rapid test results are clearly intended to provide only "preliminary" information for initial patient management until more definitive tests can be completed. Such preliminary results need to be clearly differentiated from the final, or definitive laboratory test results. In this regard, the laboratory has a responsibility for defining the confidence levels for each specific result category. Further, laboratory personnel should be sufficiently knowledgable to be able to explain the reliability of the results being provided to its physicians.

In summary, for the results of rapid automated instrument procedures to realize their full impact, the microbiology laboratory must strive to improve clinician awareness of the laboratory's capabilities to produce such results and must provide effective means to communicate them in a timely fashion which is complementary to the practice habits of physicians.

REFERENCES

1. Matsen, J., Rapid reporting of results — impact on patient, physician, and laboratory, *Rapid Methods and Automation in Microbiology,* Tilton, R. C., Ed., American Society of Microbiology, Washington, D.C., 1982, 98.
2. Maki, D. G. and Schuna, A. A., A study of antimicrobial misuse in a university hospital, *Am. J. Med. Sci.,* 275, 271, 1978.
3. Matsen, J. M., Means to facilitate physician acceptance and use of rapid test results, *Diagn. Microbiol. Infect. Dis.,* 2, 735, 1985.
4. Jorgensen, J. H., Clinical situations in which rapid methods may best be applied, *Diagn. Microbiol. Infect. Dis.,* 2, 3S, 1985.
5. Bryan, C. S., Reynolds, K. L., and Brenner, E. R., Analysis of 1,186 episodes of Gram-negative bacteremia in non-university hostpials: the effects of antimicrobial therapy, *Rev. Infect. Dis.,* 5, 629, 1983.
6. Kreger, B. E., Craven, D. E., and McCabe, W. R., Gram-negative bacteremia. IV. Re-evaluation of clinical features and treatment in 612 patients, *Am. J. Med.,* 68, 344, 1980.
7. Weinstein, M. P., Murphy, J. R., Reller, L. B., and Lichtenstein, K. A., The clinical significance of positive blood cultures: a comprehensive analysis of 500 episodes of bacteremia and fungemia in adults. II. Clinical observations, with special references to factors influencing prognosis, *Rev. Infect. Dis.,* 5, 54, 1983.
8. Doern, G. V., Clinical impact of rapid susceptibility testing, *Diagn. Microbiol. Infect. Dis.,* 3, 59S, 1985.

Future Developments

Chapter 18

FUTURE DEVELOPMENT OF AUTOMATED INSTRUMENTS FOR MICROBIOLOGY

Jack Goldstein

Clinical microbiology has undergone only the first of several phases in the establishment of automated instrumentation in the laboratory. This first phase began with a slow and skeptical acceptance of instrumented methods in the laboratory. This may have been due largely to the prevailing attitude that reading biochemical reactions utilized in microbial identifications required an experienced microbiologist. Moreover, presumably because of the retrospective nature of our science, technological developments in microbiology evolved slowly and as a result, until recently, clinical microbiologists were unexposed to rapid techniques and mechanistic approaches for the performance of microbiological analyses.

The first types of instruments widely accepted in the microbiology laboratory were those that required a microbiologist's interpretation of biochemical and antimicrobic susceptibility tests which were miniaturized versions of conventional assays.[1-2] These instruments allowed the microbiologist to input reactions manually via an apparatus that would transmit an electronic signal, signifying the outcome of a reaction, to a computer. Data bases were then accessed and a probability of identification computed automatically along with an antibiotic susceptibility profile and further interpretation in some instances. Several of these computer systems had epidemiology software which enabled the laboratorian to follow nosocomial infections based upon data accrued.[3-4] This level of automation eliminated transcription of data to work sheets, translation of numerical codes into genera and species from code books, manual compilation, and calculation of epidemiological data and general transcription errors in data transfer to patient report forms, etc.

The next level of automation added instruments that actually read biochemical reactions and turbidity in antimicrobic containing suspensions and then automatically processed these data.[1,5-9] This technique added the consistency and objectivity of instrumented spectrophotometric and turbidometric analysis to microbiological assays. Several systems, however, also allowed manual overrides to be inputed into the system by microbiologists comparing manual to machine interpretations and opting to override machine readings with visual interpretations. These instruments added some sophistication in reading results but still required constant interactive operator interface to inoculate systems, transfer them to an incubator and return them to the instrument for reading, usually one at a time.

The last level of phase I instrumentation that made its way into the clinical microbiology lab had an even slower acceptance. This type of instrument was more sophisticated than the others in terms of engineering and automation. Previously inoculated biochemicals or susceptibility tests were placed in a self-contained incubator/reader by the operator. The instrument then automatically read biochemicals and susceptibility tests and after certain criteria were met made a final reading and automatically printed out results.[10-18] This next step still required inoculation and transfer of the system into an incubator but alleviated operator interface after this step. The instrument would then automatically perform microbiological analyses either in a batch or interactive mode. Although some of the mechanical aspects of this instrument were more sophisticated than others in that a simple robotics system was utilized to move biochemical/

drug-containing panels from the incubator to an optical system, it did not incorporate any advanced liquid handling techniques to inoculate the system nor did it take advantage of new technology in its optical design or microprocessor system.

All of the automated techniques in use to date require microbial growth and are based upon miniaturized conventional microbiological tests analyzed on the basis of the production of metabolic end-products or the ability to grow in the presence or absence of various substrates or inhibitory substances. Susceptibility testing is based solely upon turbidometric or nephalometric analysis after sufficient time required for the development of optical densities detectable by the photosensors utilized in each system.

After such a slow evolution and transition from no automation through phase I, what will occur in phase II? Phase II has already begun. It includes more automated techniques to detect conventional microbiological reactions using new liquid-handling techniques and robotics to inoculate systems, add reagents and move inoculated systems from one station to the next for automatic processing. More elaborate optical systems using fiber optics, more sensitive photodetectors, bar code readers and more complex computers will allow more hands-off microbiology and more same-day identification and susceptibility using growth-based systems now in existence. Additionally, new identification and susceptibility systems still requiring organism inocula from a solid medium but not requiring growth for identification and only a minimal number of generations of bacterial growth to predict susceptibility to antimicrobial agents must also be considered. The reagent systems, many of which have already entered the market in the past 3 to 4 years utilize the hydrolysis of enzymatic substrates to profile and identify bacteria in 3 to 5 hr after inoculation with a turbid suspension of microorganisms from plated media.[19-20] Moreover, antimicrobial susceptibilty tests have been developed which detect the elaboration of enzymes which inhibit certain antibiotics[21] or which predict susceptibility or resistance based upon only several generations of growth,[1] reduction of tetrazolium chloride,[22] or production of ATP indicative of metabolic activity.[23] Most of these methods, except ATP detection, are applicable to instruments developed in phase I and II using carriers with multiple compartments and performing simultaneous analysis of compartments at various intervals after inoculation. These reactions include colorimetric, turbidimetric, or fluorometric determinations handled in an automated fashion to effect a 3 to 5 hr identification and susceptibilty while retaining the accuracy of more conventional methods and remaining within the realm of experience of most microbiologists. ATP detection, of course, is performed using luminescent technology and can be performed in a semi-automated fashion using a continuous flow-type system.[24] Added to the level of sophistication will probably be new mathematical models to identify bacteria not based solely upon a positive percentage chart and calculated probabilities of correct identifications.

The real future of automation lies in new techniques that can address organism identifications and susceptibilities directly in patient specimens. Only a few reagent systems and/or instrumented approaches seem to have the potential for identifying microorganisms and/or performing antimicrobial susceptibility tests. Most of these have grown out of new technologies that have recently become available and most, in addition to being applicable to bacteria, have the potential for identifying viruses, parasites, etc. and performing antimicrobial assays. Although the list is somewhat more extensive than those listed, the major systems with probable usefulness in phase III are the use of monoclonal antibodies and DNA probes in conjunction with automated phase II instrumentation capable of reading both color and fluorescence, flow cytometric assay in conjunction with various types of lasers for detection of light scattering patterns obtained with nuclear and other stains, circular intensity differential light scattering, mass spectrometry, nuclear magnetic resonance spectrometry, and imaging analysis using lasers in conjunction with a variety of reagent systems.

Hybridoma technology resulting in the production of monoclonal antibodies offers new opportunities through the use of highly specific antibodies for the rapid diagnosis of infectious etiologic agents. Antibodies produced by this technique are uniform, highly specific, easy to manufacture and applicable to automation with phase I and II instrumentation. Monoclonal-based diagnostics for microbial antigens can be assayed using nephalometric, colorimetric, or fluorometric measurements either one test at a time or in an array containing multiple wells measuring individual tests or several dilutions of the same test. DNA hybridization techniques also are advantageous for infectious disease assays. They have the potential for a high degree of specificity among closely related organisms (such as the Enterobactereacae) that cannot be differentiated by immunochemical means. At present, the principal disadvantage of this approach is the means of detecting hybridization. Most DNA probe assays utilize radioactive isotopes for detection.[25] Several methods have been developed which utilize fluorescent detection systems.[25] One instrument already manufactured utilizes microdilution technology, a robotics system for moving panels from station to station, an automated reagent/liquid-dispensing apparatus capable of dispensing several liquids simultaneously to a row of micro-wells, the ability to wash and aspirate fluids from wells, a fluorescence optical system, and a microprocessor to capture, analyze, and interpret data.[26] Other instruments already utilized for drug monitoring can also be used in a qualitative type of infectious disease assay. One such instrument dilutes specimens, adds reagents, and reads reactions by means of a fluorescence polarization analyzer in a fully automated fashion.[27] The use of new direct monoclonal or DNA probe tests will permit a more rapid diagnosis of infectious agents that may allow for earlier treatment of patients with appropriate medication. This may have profound implications in light of DRGs in that the use of these assays may be translated into shorter patient stays and reduced costs.[28] Direct detection systems will also allow diagnosis of organisms difficult to grow and identify in the clinical microbiology laboratory.

Further in the future is the possible use of a flow cytometer for microbial identification and susceptibility testing. Flow cytometry can be utilized to make accurate measurements of fluorescence emitted from cells in seconds. Cells to be analyzed are injected into the center of a stream of fluid and pass single file through an intense light beam, usually from a laser.[29] Cells passing through the beam scatter light. The amount of scattered light gives an indication of the size of the cells. Fluorescent dyes in or on the cells are excited by the illuminating beam and fluorescence emitted from cells is collected by a lens and converted into an electrical signal by a photodetector and associated electronics. Using this basic technique several investigators have developed applications for microbial identification and susceptibility. Work on bacterial systems has been based upon DNA content comparisons using: stains such as ethidium bromide and mitomycin C; RNA content; protein content to cell size, and GC ratio analyses.[30-33] Work with susceptibility has been accomplished through comparisons of total DNA content in the presence and absence of antimicrobial substances.[33] The flow cytometric technique has the capacity to look at small numbers of organisms, mixed populations of organisms, organisms directly from specimens, and is applicable to bacteria, viruses, and protozoa, as well as other microorganisms. Several detection systems can be utilized in conjunction with a cytometer and thus differentiation of organisms is not limited solely to the use of nuclear and cytoplasmic stains and dyes utilized to date. Most identifications using this system could be performed within minutes, susceptibilities predicted within 2 or 3 generations for most cidal drugs. Some drawbacks of this technology include problems with alignment of currently manufactured instruments and clogging of the cell stream. In addition, the use of most laser systems results in a final instrument too costly for routine clinical use in most hospital or reference microbiology laboratories. It is also difficult to conceptualize how multiple antibiotics and specimens could be handled efficiently using this technique.

A related technique sometimes utilized in conjunction with a flow cytometer is called CIDS or circular intensity differential light scattering.[34] This refers to the differential scattering of left and right circularly polarized light and is the amount of light scattered when an incident beam is left circularly polarized minus the amount scattered with the incident beam is right circularly polarized, divided by the total amount of scattered light.[34] It has been theorized that CIDS is sensitive to differences in chiral structures and thus can differentiate organisms on the basis of differences in the supercoil structure of DNA.[35] Experiments have demonstrated the use of this technique to distinguish among different species and serotypes within the genus *Legionella*.[36] It has been postulated that by coupling CIDS with a flow cytometer, mixed populations of organisms containing both bacteria and viruses can be identified as they pass individually in the cytometer, through the laser beam.[34,36] The challenge with CIDS is to prove its ability to differentiate multiple species of microorganisms and to produce a CIDS-based instrument affordable by the clinical microbiology laboratory.

Different applications for microbial identification have been the subject of many investigations involving gas chromatography,[37] nuclear magnetic resonance spectrometry, and mass spectrometry.[38] Recently, a group of British investigators have presented data on the applicability of pyrolysis mass spectrometry for microbial identification.[39] Pyrolysis mass spectrometry involves the thermal degradation of organic material by heating bacteria to its Curie point, resulting in the production of a mixture of volatile pyrolysis products. Thermal degradation is carried out in close proximity to the ion source of a mass spectrometer, and the resultant pyrolysis products are ionized and detected as a mass spectrum fingerprint. Quantitative information is captured by a computer which performs multivariate analysis and can compare information derived from an experimental isolate to a data base comprised of knowns. Mass spectrometry can be utilized to identify bacteria in less than 10 min with relatively little sample preparation. The type of system utilized is amenable to automation and computerization. This technique is in the very early stages of development but may have long term potential in diagnostic microbiology. Again some of the present drawbacks include the complexity of the methods which bridge microbiology, analytical chemistry, and complex mathematical modeling. Most clinical microbiologists have little training or experience in these other disciplines and have never operated a mass spectrometer. The development of an instrument for this purpose, including a microprocessor fast enough to capture data from a mass spectrometer and capable of analyzing the data is, at present, quite expensive.

The last technology with potential for performing microbial analysis in the future is imaging analysis in conjunction with one or several reagent or staining systems. Several research instruments have been developed which can differentiate cells into segregated populations based upon chemical or physical properties.[40] Some can accomplish this from a slide or plate via a laser-illuminated microscopic system equipped with computer-controlled beam and stage.[41] A microprocessor is also required for comparing data output and analysis. Utilizing this type of technology, large surface areas can automatically be scanned with a laser after staining or performing biochemical reactions with fluorescence indicators. In this manner, similar measurements as those taken with a flow cytometer can be made with an imaging system. This technology represents a potentially powerful automated analytical method for studying structural or biochemical parameters of bacterial cells.

The object of this chapter was to predict the future of automation in clinical microbiology. In order to accomplish this task, it was essential to look back into the past and base the future, in part, on the progression of automation that has taken place over the past decade. Also important to consider are new technology-driven products based on hybridoma and DNA probe systems, advances in computer and laser technology, and lastly, on our own experience as microbiologists in looking at improvements needed in current instrumentation. Developments in liquid handling and automated techniques for conventional methodologies have already begun. Costs of microprocessors and lasers have declined over the past several years and will probably continue to decline with new advances. The next technology to emerge will utilize one or a combination of those described and perform microbial analyses automatically, directly from patient samples within minutes so that specific diagnoses can be made and therapy prescribed bringing clinical microbiology into the realm of proactive medicine instead of a retrospective science.

REFERENCES

1. Thornsberry, C. and Gavan, T. L., Automated procedures for antimicrobial susceptibility tests, in *Manual of Clinical Microbiology,* 3rd ed., Lennette, E. H., Balows, A., Hausler, W. J., and Truant, J. P., Eds., American Society for Microbiology, Washington, D.C., 1980, chap. 48.
2. Operator's Manual, *touchSCAN-IBM,* American Micro Scan, Sacramento, 1983, 1.1—7.3.
3. Ryan, R. J. and Peebles, J. E., On-line computer entry of routine and AutoMicrobic system bacteriology results, in *Rapid Methods and Automation in Microbiology,* Tilton, R. C., Ed., American Society for Microbiology, Washington, D.C., 1981, 23.
4. Wertz, R., autoScan data management system, in *Rapid Methods and Automation in Microbiology,* Tilton, R. C., Ed., American Society for Microbiology, Washington, D.C., 1981, 30.
5. Barnes, W. G., Green, L. R., and Talley, R. L., Clinical evaluation of automated antibiotic susceptibility testing with the MS-2 system, *J. Clin. Microbiol.,* 12, 527, 1980.
6. Hansen, S. L. and Freedy, P. K., Concurrent comparability of automated systems and commercially prepared microdilution trays for susceptibility testing, *J. Clin. Microbiol.,* 17, 878, 1983.
7. Jones, R. N., Thornsberry, C., Barry, A. L., and Gavan, T. L., Evaluation of the Sceptor microdilution antibiotic susceptibility testing system: a collaborative investigation, *J. Clin. Microbiol.,* 13, 184, 1981.
8. Rhoden, D. L., Baker, C. N., Stocker, S., and Smith, P. B., Identification of clinically significant bacteria with the autoSCAN-4, presented at Int. Symp. Rapid Methods and Automation in Microbiology and Immunology, Berlin, June 6 to 9, 1984, P89.
9. Thornsberry, C., Anhalt, J. P., Washington, J. A., II, McCarthy, L. R., Schoenknecht, F. D., Sherris, J. C., and Spencer, H. J., Clinical laboratory evaluation of the Abbott MS-2 automated antimicrobial susceptibility testing system: report of a collaborative study, *J. Clin. Microbiol.,* 12, 375, 1980.
10. Aldridge, C., Jones, P. W., Gibson, S., Lanham, J., Meyer, M., Vannest, R., and Charles, R., Automated microbiological detection/identification system, *J. Clin. Microbiol.,* 6, 406, 1977.
11. Barry, A. L. and Badal, R. E., Identification of Enterobacteriaceae by the AutoMicrobic system: Enterobacteriaceae biochemical cards versus Enterobacteriaceae-plus biochemical cards, *J. Clin. Microbiol.,* 15, 575, 1982.
12. Dudley, M., Nauschuetz, W. F., and Juchau, S. V., Clinical evaluation of new AMS GSC plus card, *Diagn. Microbiol. Infect. Dis.,* 1, 139, 1983.
13. Goldstein, J., Guarneri, J. J., Della-Latta, P., and Scherer, J., Use of the AutoMicrobic and Enteric-Tek systems for identification of Enterobacteriaceae, *J. Clin. Microbiol.,* 15, 654, 1982.
14. Goldstein, J., Guarneri, J. J., and Scherer, J., Use of the AutoMicrobic system for rapid antimicrobial susceptibility testing of Enterobacteriaceae in a clinical laboratory, *J. Clin. Lab. Automat.,* 2, 329, 1982.
15. Johnson, J. E., Jorgensen, J. H., Crawford, S. A., Redding, J. S., and Pruneda, R. C., Comparison of two automated instrument systems for rapid susceptibility testing of gram-negative bacilli, *J. Clin. Microbiol.,* 18, 1301, 1983.

16. Kelly, M. T., Latimer, J. M., and Balfour, L. C., Comparison of three automated systems for antimicrobial susceptibility testing of gram-negative bacilli, *J. Clin. Microbiol.*, 15, 902, 1982.
17. Smith, P. B., Gavan, T. L., Isenberg, H. D., Sonnenwirth, A., Taylor, W. I., Washington, J. A., II, and Balows, A., Multi-laboratory evaluation of an automated microbial detection/identification system, *J. Clin. Microbiol.*, 8, 657, 1978.
18. VanHorn, K. G., Vandernoot, A. M., Burke, E. W., and McKitrick, J. C., Evaluation of the AutoMicrobic system gram-negative general susceptibility-plus card, *J. Clin. Microbiol.*, 20, 630, 1984.
19. D'Amato, R. F., Eriquez, L., Tomfohrde, K. M., and Singerman, E., Rapid identification of *Neisseria gonorrhoeae* and *Neisseria meningitidis* by using enzymatic profiles, *J. Clin. Microbiol.*, 7, 77, 1978.
20. Humble, M. W., King, A., and Phillips, I., API ZYM: a single rapid system for the detection of bacterial enzymes, *J. Clin. Pathol.*, 30, 275, 1977.
21. Anhalt, J. P., Sabath, L. D., and Barry, A. L.,Bactericidal activity, activity of antimicrobics in combination and detection of beta-lactamase production, in *Manual of Clinical Microbiology*, 3rd ed., Lennette, E. H., Balows, A., Hausler, W. J., and Truant, J. P., Eds., American Society for Microbiology, Washington, D. C., 1980, chap. 46.
22. Bartlett, R. C. and Mazens, M. F., Rapid antimicrobial susceptibility test using tetrazolium reduction, *Antimicrob. Agents Chemother.*, 15, 769, 1979.
23. Harber, M. J. and Asscher, A. W., A new method for antibiotic assay based on measurement of bacterial ATP using the firefly bioluminescent system, *Antimicrob. Agents Chemother.*, 3, 35, 1977.
24. Whitehead, T. P., Kricka, L. J., Carter, J. N., and Thorpe, G. H. G., Analytical luminescence: its potential in the clinical laboratory, *Clin. Chem.*, 25, 1531, 1979.
25. Falkow, S. and Moseley, S. L., Specific Probes in Diagnostic Microbiology, U.S. Patent 4, 358, 535, 1982.
26. Jolley, M. E., Wang, C.J., Ekenberg, S. J., Zuelke, M. S., and Kelso, D. M., Particle concentration fluorescence immunoassay (PCFIA): a new, rapid immunoassay technique with high sensitivity, *J. Immunol. Methods*, 67, 21, 1984.
27. Jolley, M. E., Stroupe, S. D., Schwenzer, K. S., Wang, C. J., Lusteffes, M., Popelka, S. R., Holen, J. T., and Kelso, D. M., Fluorescence polarization immunoassay. III. An automated system for therapeutic drug determination, *Clin. Chem.*, 27, 1575, 1981.
28. Drake, P. F., *Biomedical Technology Industry Analysis: — Monoclonal Antibodies and DNA Probes*, Kidder, Peabody, New York, 1984, 1—36.
29. Pinkle, D., Flow cytometry and sorting, *Anal. Chem.*, 54, 503A, 1982.
30. Langlois, R., VanDilla, M. A., Yajko, D., and Hadley, W. K., Rapid characterization of bacteria in culture and clinical specimens by the use of flow cytometry, presented at Annual Meeting of the American Society for Microbiology, New Orleans, March 6—11, 1983, C251.
31. Steen, H. B. and Boye, E., *Escherichia coli* growth studied by dual parameter flow cytometry, *J. Bacteriol.*, 145, 1091, 1980.
32. Steen, H. B. and Boye, E., Bacterial growth studied by flow cytometry, *Cytometry*, 1, 32, 1980.
33. Boye, E., Skarstad, K., and Steen, H. B., Flow cytometry in clinical microbiology, presented at Annual Meeting of the American Society for Microbiology, New Orleans, March 6—11, 1983, C252.
34. Saltzman, G. C. and Gregg, C. T., Current and experimental methods of rapid microbial identification. Circular intensity differential light scattering, *Biotechnology*, March 1984, p. 243.
35. Bustamante, C., Maestre, M. F., and Tinoco, I., Circular intensity differential scattering of light by helical structures. I. Theory, *J. Chem. Phys.*, 73, 4273, 1980.
36. Gregg, C. T., Grace, W. R., and Saltzman, G. C., Rapid identification of microorganisms by circular intensity differential light scattering, presented at Annual Meeting of the American Society for Microbiology, New Orleans, March 6—11, 1983, C249.
37. Edman, D. C., Craven, R. B., and Brooks, J. B., Gas chromatography in the identification of microorganisms and diagnosis of infectious diseases, *Crit. Rev. Clin. Lab. Sci.*, 14, 133, 1981.
38. Ruterjans, H., NMR in the study of cells and tissues, presented at Int. Symp. Rapid Methods and Automation in Microbiol. Immunol., Berlin, June 6—9, 1984, S79.
39. Shute, L. A., Gutteridge, C. S., Norris, J. R., and Berkeley, R. C. W., Curie-point pyrolysis mass spectrometry applied to characterization and identification of selected *Bacillus* species, *J. Gen. Microbiol.*, 130, 343, 1984.
40. Schindler, M., Osborn, M. J., and Koppel, D. E., Lateral diffusion of lipopolysaccharide in the outer membrane of *Salmonella typhimurium*, *Nature (London)*, 285, 261, 1980.

Index

INDEX

A

B

E

F

G

L

M

N

S

T